iPhone®
For Seniors

2023 Edition

by Dwight Spivey

iPhone® For Seniors For Dummies®, 2023 Edition

Published by: John Wiley & Sons, Inc., 111 River Street, Hoboken, NJ 07030-5774, www.wiley.com

Copyright © 2023 by John Wiley & Sons, Inc., Hoboken, New Jersey

Published simultaneously in Canada

For general information on our other products and services, please contact our Customer Care Department within the U.S. at 877-762-2974, outside the U.S. at 317-572-3993, or fax 317-572-4002. For technical support, please visit https://hub.wiley.com/community/support/dummies.

Wiley publishes in a variety of print and electronic formats and by print-on-demand. Some material included with standard print versions of this book may not be included in e-books or in print-on-demand. If this book refers to media such as a CD or DVD that is not included in the version you purchased, you may download this material at http://booksupport.wiley.com. For more information about Wiley products, visit www.wiley.com.

Library of Congress Control Number: 2022917514

ISBN 978-1-119-91284-2 (pbk); ISBN 978-1-119-91285-9 (ebk); ISBN 978-1-119-91286-6

SKY10036288_101022

Table of Contents

Introduction

Apple's iPhone is designed to be easy to use, but you can still spend hours exploring the preinstalled apps, discovering how to change settings, and figuring out how to sync the device to your computer or through iCloud. (If you don't know what iCloud is, no worries; we dive into it in Chapter 4.) I've invested those hours so that you don't have to — and I've added battle-tested advice and tips so that you can become an expert with your iPhone, regardless of which model you own.

This book will get you up and running with your iPhone quickly and painlessly so that you can confidently move on to the fun part. Apple does a legendary job of making its devices and software intuitive, but owning this book is akin to having a good friend by your side who's tech-savvy and ready to lend a helping hand when you need it.

About This Book

This book is written for mature people like you — folks who may be relatively new to using a smartphone and who want to discover the basics of buying an iPhone, making and receiving phone and video calls, working with apps, getting on the internet, enjoying music and photos, and discovering all the other tricks the iPhone can do. In writing this book, I've tried to consider the types of activities that might interest someone who is 50 years old or older and picking up an iPhone for the first time. As a quinquagenarian myself, I want to make certain that you get the most bang for your buck with this tome.

Foolish Assumptions

This book is organized by sets of tasks. These tasks start at the beginning, assuming that you've never laid your hands on an iPhone, and guide you through basic steps using nontechnical language.

iPhone For Seniors For Dummies covers going online using either a Wi-Fi or cellular connection, browsing the web (Chapter 12), and checking email (Chapter 13). I also assume that you'll want to use the Apple Books e-reader app, so I cover its features in Chapter 17. I tackle all kinds of neat things you can do with your iPhone, such as customizing your experience with iOS 16's new approach to lock screens (Chapter 2), monitoring the use of your iPhone and its apps (Chapter 5), discovering new apps (Chapter 14), taking and sharing photos and videos (Chapters 19 and 20, respectively), and tracking your health (Chapter 24)!

Icons Used in This Book

Icons are tiny pictures in the margin that call your attention to special advice or information.

TIP

This brief piece of advice helps you take a skill further or provides an alternate way to do something.

WARNING

Heads up! This is something that might wreak havoc on your iPhone or that could be difficult or expensive to undo.

REMEMBER

This information is so useful, it's worth keeping in your head — not just on your bookshelf.

TECHNICAL STUFF

This information isn't essential, but it's neat to know.

Beyond the Book

There's even more iPhone information on www.dummies.com. This book's cheat sheet offers tips on using Siri and suggests all kinds of apps that you can use to make the most of your iPhone. To get to the cheat sheet, go to www.dummies.com, and then type **iPhone For Seniors For Dummies Cheat Sheet** in the search box.

Where to Go from Here

You can work through this book from beginning to end or simply open a chapter to solve a problem or acquire a specific new skill. The steps in each task quickly get you where you want to go, without a lot of technical explanation.

When I wrote this book, all the information was accurate for the iPhone SE (second generation or later), 8 and 8 Plus, X (the Roman numeral for ten), XR, XS, XS Max, 11, 11 Pro, 11 Pro Max, 12, 12 mini, 12 Pro, 12 Pro Max, 13, 13 mini, 13 Pro, 13 Pro Max, 14, 14 Plus, 14 Pro, and 14 Pro Max, along with version 16 of iOS (the operating system used by the iPhone).

Apple is likely to introduce new iPhone models and versions of iOS between book editions. If you've bought a new iPhone and found that its hardware, user interface, or iPhone-related software on your computer (such as iTunes or Music) looks a little different than what is presented here, check out what Apple has to say at www.apple.com/iphone and www.apple.com/ios. You'll find updates on those sites regarding the company's latest releases.

1

Getting to Know Your iPhone

IN THIS CHAPTER

» **Discover what's new in iPhones and iOS 16**

» **Choose the right iPhone for you and find where to buy it**

» **Understand what you need to use your iPhone**

» **Explore what's in the box**

» **Take a look at the gadget**

Chapter **1**

Buying Your iPhone

You've read about it. You've seen the lines at Apple Stores on the day a new version of the iPhone is released. You're so intrigued that you've decided to get your own iPhone so you can have a smartphone that can do much more than make and receive calls. Perhaps you're not new to smartphones but are ready to make the switch to the ultimate in such devices. With your iPhone, you can have fun with apps such as games and exercise trackers; explore the online world; read e-books, magazines, and other periodicals; take and organize photos and videos; listen to music and watch movies; and a lot more.

Trust me: You've made a good decision, because the iPhone redefines the mobile phone experience in an exciting way. It's also a perfect fit for seniors.

In this chapter, you learn about the advantages of the iPhone, as well as where to buy this little gem and associated data plans. After you have one in your hands, I help you explore what's in the box and get an overview of the little buttons and slots you'll encounter — luckily, the iPhone has very few of them.

Discover the Newest iPhones and iOS 16

Apple's iPhone gets its features from a combination of hardware and its software operating system, which is called *iOS* (short for *iPhone operating system*). The most current version of the operating system is iOS 16. It's helpful to understand which new features the latest models and iOS 16 bring to the table (all of which are covered in more detail in this book).

Apple's latest additions to the iPhone family are the iPhone 14, 14 Plus, 14 Pro, and 14 Pro Max. Like their predecessors, they're highly advanced smartphones that leave competitors in the dust. Here are some of the key features of the latest iPhone models:

» **A15 and A16 Bionic chips:** The iPhone 14 and 14 Plus models include the A15 chip, while the 14 Pro and 14 Pro Max receive the latest and greatest, the A16. The truly innovative tech in these models demands processors that can handle some heavy lifting while still being able to answer calls and retrieve email, and the A15 and A16 are both more than capable.

» **Dynamic Island (14 Pro and Pro Max models only):** No, this isn't the latest reality show craze, but rather a long-awaited innovation. iPhone models of late have sported a notch at the top of their screens where sensors, cameras, and other hardware reside. Dynamic Island is a seamless pairing of hardware and software that effectively makes that area come alive with information for you, making it an upgraded notch with a twist, if you will. The notch appears to expand or contract, depending on the notifications, alerts, and other activities its currently tasked with. I hope this great update will find its way into other iPhone models moving forward.

» **Emergency SOS via Satellite and Crash Detection:** Apple has incorporated these two critically important new safety features into the iPhone 14 lineup. Emergency SOS via Satellite helps you reach emergency responders when you're outside traditional cell or Wi-Fi service. Crash Detection utilizes new gyroscope and accelerometer tech in the latest models to detect when you've been in an automobile crash and will cause your iPhone to alert emergency services automatically. Both features are something you never want to need but are thankful to have.

» **Splash, water, and dust resistance:** Your new iPhone 14, 14 Plus, 14 Pro, or 14 Pro Max is resistant to damage caused by water splashing onto it or from dust collecting in it. Now, you don't want to take your iPhone 14 model deep-sea diving, but it's likely to survive submersion in about six meters of water for up to 30 minutes. In other words, if your iPhone 14 model gets wet, it's much more likely to survive the ordeal than older iPhone iterations, but it still isn't something you'd like to see happen to your expensive investment.

TIP

You might consider acquiring AppleCare+, which is Apple's extended warranty, currently priced at $149 (iPhone 14), $179 (14 Plus), or $199 (iPhone 14 Pro and 14 Pro Max) per year. Monthly plans are also available. AppleCare+ covers unlimited incidents of accidental damage (but you will be charged minimal fees, based on the nature of the repair), which could more than cover the cost of repairing your iPhone without it. You can also get AppleCare+ with theft and loss coverage for an additional $70 (all four models).

» **Ceramic Shield:** The toughness and durability of Apple's screens just keeps getting better. Ceramic Shield was developed by Apple and Corning, and according to them, it's the toughest screen ever for a smartphone, making it four times more likely than other smartphones to survive a drop unscathed.

WARNING

Don't think your iPhone is unbreakable. Cases are still a good — no, make that a great — idea. As mentioned, Apple has a line of cases that not only protect your iPhone but also allow for wireless MagSafe and Qi charging. (Qi is an industry-standard wireless charging technology used by Apple and most smartphone manufacturers.)

Any iPhone model from the iPhone 8 and newer (including the SE second generation, and all 11, 12, 13, and 14 models) can use most features of iOS 16 if you update the operating system (discussed in detail in Chapter 3). This book is based on iOS 16. This update to the operating system adds many features, including (but definitely not limited to) the following:

» **All-new lock screen:** Apple's taken a fresh approach to the lock screen, allowing you to customize it to your heart's content. You can create lock screens for every occasion, switch between them in a snap, and include items like widgets, live activities, weather, and more.

» **Focus:** Think of Focus as an extension of the Do Not Disturb feature. You can customize a focus to filter notifications based on what you're doing at the moment. iOS 16 introduces new features like focus schedules, focus filters (imagine one for work and one for personal), and allow and silence lists for apps and contacts.

» **Photos:** Photos in iOS 16 includes the new iCloud Shared Photo Library feature, which allows you to create a library of photos that you can share with others via iCloud. Other participants may also collaborate by adding their own photos to the library, providing a more complete memory experience for all. Everyone can also edit, delete, caption, and mark as a favorite any photo in the library.

» **Messages:** The latest iteration of Messages finally allows you to select multiple messages at once (for example, if you want to delete several at one time), mark read messages as unread, edit messages you've already sent (up to 15 minutes after), and more. This is a nice upgrade, IMO.

» **Safari:** Tab groups, a welcome new feature in iOS 15, allow you to group your open web pages any way you like. iOS 16 takes the feature a step further by allowing you to share tab groups and create pinned tabs in tab groups. It also introduces Passkeys, a new and more secure way to authenticate yourself on websites that require a password.

» **Maps:** Maps now allows you to add multiple stops along your route. The new Transit Fares feature helps you calculate fares and other fees so you can better prepare for trip costs.

>> **Health app:** You can now use Health to track your medications, discover potential interaction issues, add medications by scanning the label on bottles, get reminders when it's time to take medications, and more.

These are but a few of the improvements made to the latest version of iOS. I suggest visiting www.apple.com/ios/ios-16 to find out more.

TIP

Don't need all the built-in apps? You can remove them from your Home screen. When you remove a built-in app from your Home screen, you aren't deleting it — you're hiding it. (Note that built-in apps take up very little of your iPhone's storage space.) And if you change your mind, you can easily add them back to your Home screen by searching for them in the App Store and tapping the Get button, or by retrieving them from the App Library. How you recover them depends on the app; some allow you to hide them while others only let you relegate them to the App Library.

Choose the Right iPhone for You

A variety of iPhone models are on the market; it can be daunting when trying to decide which one you want to purchase. In this section, I focus on Apple's newest models, the iPhone 14 series. If you'd like to explore others, Apple has a great tool for making comparisons at www.apple.com/iphone/compare.

The sizes of the latest iPhone 14 models vary:

>> iPhone 14 measures 2.82" by 5.78" (6.1" diagonally) with a depth of .31 inch (see **Figure 1-1**).

>> iPhone 14 Plus measures 3.07" by 6.33" (6.7" diagonally) with a depth of .31 inch (also shown in Figure 1-1).

>> iPhone 14 Pro measures 2.81" by 5.81" (6.1" diagonally) with a depth of .31 inch (see **Figure 1-2**).

>> iPhone 14 Pro Max measures 3.05" by 6.33" (6.7" diagonally) with a depth of .31 inch (also shown in Figure 1-2).

Image courtesy of Apple, Inc.

FIGURE 1-1

Image courtesy of Apple, Inc.

FIGURE 1-2

You can get iPhone 14 and 14 Plus in starlight, midnight, blue, purple, and a beautiful product red version. iPhone 14 Pro and 14 Pro Max come in gold, silver, space black, and a great-looking deep purple.

Not sure whether to get an iPhone 14 model? Here are a few more key differences:

» **All iPhone 14 models include upgraded batteries.** You get up to 20 hours of video playback for iPhone 14, up to 23 hours for 14 Pro, 26 hours for 14 Plus, and a whopping 29 hours for 14 Pro Max.

» **iPhone 14 models use eSIMs instead of physical SIMs.** A SIM stores important information about your phone and your cellular provider network. Physical SIMs can be moved from phone to phone, but eSIMs are permanently built-in.

» **All models received camera upgrades.** iPhone 14 Pro and 14 Pro Max have triple rear-facing cameras, providing amazing optical zoom, portrait mode, and other features. The 14 and 14 Plus have dual rear-facing cameras.

» **Screen resolution.** The higher the resolution the better, especially for larger screens since you need to pack more pixels (the tiny dots of color that make up the images) into a larger space. The iPhone 14 offers 2532 x 1170 resolution; 14 Plus provides 2778 x 1284 resolution; 14 Pro boasts 2556 x 1179; and 14 Pro Max provides a stunning 2796 x 1290.

Table 1-1 gives you a quick comparison of the iPhone SE (third generation), 12, 13, 13 mini, 14, 14 Plus, 14 Pro, and 14 Pro Max (models currently sold by Apple). All costs are as of the time this book was written. (Some carriers may introduce non-contract terms.)

Other differences between iPhone models come primarily from the current operating system, iOS 16, which I cover in the remaining chapters of the book.

One exciting pricing option is the iPhone Upgrade Program. You choose your carrier, get an unlocked phone so you can change carriers, and receive Apple Care+ to cover you in case your phone has problems, all starting at a cost of $39.50 a month (depending on the iPhone model you select). The price does not include data usage from your carrier. Check out `www.apple.com/shop/iphone/iphone-upgrade-program` for more information.

TABLE 1-1 **iPhone Model Comparison**

Model	Storage	Cost (may vary by carrier)	Carriers
SE (third generation)	64GB, 128GB, and 256GB	From $429	AT&T, Verizon, Sprint, T-Mobile
12	64GB, 128GB, and 256GB	From $599	AT&T, Verizon, Sprint, T-Mobile
13	128GB, 256GB, and 512GB	From $699	AT&T, Verizon, Sprint, T-Mobile
13 mini	128GB, 256GB, and 512GB	From $599	AT&T, Verizon, Sprint, T-Mobile
14	128GB, 256GB, and 512GB	From $799	AT&T, Verizon, Sprint, T-Mobile
14 Plus	128GB, 256GB, and 512GB	From $899	AT&T, Verizon, Sprint, T-Mobile
14 Pro	128GB, 256GB, 512GB, and 1TB	From $999	AT&T, Verizon, Sprint, T-Mobile
14 Pro Max	128GB, 256GB, 512GB, and 1TB	From $1,099	AT&T, Verizon, Sprint, T-Mobile

Decide How Much Storage Is Enough

Storage is a measure of how much information — for example, movies, photos, and software applications (apps) — you can store on a computing device. Storage can also affect your iPhone's performance when handling such tasks as streaming favorite TV shows from the web or downloading music.

Streaming refers to playing video or music content from the web (or from other devices) rather than playing a file stored on your iPhone. You can enjoy a lot of material online without ever downloading its full content to your phone — and given that the most storage-endowed iPhone model has a relatively small amount of storage, streaming is a good idea. See Chapters 18 and 20 for more about getting your music and movies online.

Your storage options with an iPhone 14 or 14 Plus are 128, 256, and 512 gigabytes (GB), while 14 Pro and 14 Pro Max have 128GB, 256GB, 512GB, and 1TB (terabyte, which is 1000GB). You must choose the right amount of storage because you can't open the unit and add more, as you usually can with a desktop computer. However, Apple has thoughtfully provided iCloud, a service you can use to back up content to the internet. (You can read more about iCloud in Chapter 4.)

How much storage is enough for your iPhone? Here are some guidelines:

» If you simply want to check email, browse the web, and keep your calendar up to date, and you enjoy communicating via voice, video, and instant messaging, 128GB likely is plenty.

» For most people who manage a reasonable number of photos, download some music, and watch heavy-duty media such as movies online, 256GB may be sufficient. But if you might take things up a notch regarding media consumption and creation in the future (such as the newest grandchild being on the way soon), you should seriously consider 512GB.

» If you like lots of media, such as movies or TV shows, you might need 512GB or 1TB. For example, shooting 4K video at 60 frames per second will take roughly 1GB of storage space for every two and a half minutes of footage. If you shoot a lot of video, 1TB of storage might be more appealing.

TECHNICAL STUFF

Do you know how big a *gigabyte* (GB) is? Consider this: Just about any computer you buy today comes with a minimum of 256GB of storage. Computers have to tackle larger tasks than iPhones, so that number makes sense. The iPhone, which uses a technology called *flash storage* for storing data, is meant (to a great extent) to help you experience online media and email; it doesn't have to store much since it pulls lots of content from the internet. In the world of storage, 64GB for any kind of storage is puny if you keep lots of content (such as audio, video, and photos) on the device.

What's the price for larger storage? For the iPhone 14, a 128GB unit costs $799, 256GB is $899, and 512GB will set you back $1,099. iPhone 14 Plus with 128GB goes for $899, 256GB at $899, and 512GB for $1,199. iPhone 14 Pro with 128GB is $999, 256GB is $1,099, 512GB goes for $1,299, and the model tops out at $1,499 for 1TB. Not to be outdone, iPhone 14 Pro Max is the priciest: $1,099 for 128GB, $1,199 for 256GB, $1,399 for 512GB, and $1,599 for 1TB. Note that prices may vary by carrier and where you buy your phone.

Understand What You Need to Use Your iPhone

Before you head off to buy your iPhone, you should know what other connections and accounts you'll need to work with it optimally.

At a minimum, to make standard cellular phone calls, you need to have a service plan with a cellular carrier (such as AT&T or Verizon), as well as a data plan that supports iPhone. The data plan allows you to exchange information (such as emails and text messages) over the internet and download content (such as movies and music) without need of a Wi-Fi connection. Before you sign up, try to verify the strength of cellular coverage in your area (ask your local cellular company representatives for more info), as well as how much data your plan provides each month.

You also need to be able to update the iPhone operating system (iOS). Without a phone carrier service plan, you can update iOS either wirelessly over a Wi-Fi network or by plugging your iPhone into your computer. You would also need to use a Wi-Fi network to go online and make calls using an internet service, such as FaceTime or Skype.

TIP

Given the cost and high-tech nature of the iPhone, having to jury-rig these basic functions doesn't make much sense. Trust me: Get an account and data plan with your phone service provider.

You should also open a free iCloud account, Apple's online storage and syncing service, to store and share content online among your Apple devices. For example, you can set up iCloud in such a way that photos you take on your iPhone will appear on your iPad. You can also use a computer to download photos, music, books, or videos and transfer them to your iPhone through a process called syncing.

Apple has set up its software and the iCloud service to give you two ways to manage content for your iPhone — including apps, music, or photos you've downloaded — and specify how to sync your calendar and contact information.

There are a lot of tech terms to absorb here (iCloud, syncing, and so on). Don't worry. Chapters 3 and 4 cover those settings in more detail.

Where to Buy Your iPhone

You can't buy an iPhone from just any retail store. You can buy an iPhone at the brick-and-mortar or online Apple Store and from mobile phone providers, such as AT&T, Sprint, T-Mobile, and Verizon. You can also find an iPhone at major retailers, such as Best Buy and Walmart, through which you have to buy a service contract for the phone carrier of your choice. You can also find iPhones at several online retailers (such as Amazon.com and Newegg.com) and through smaller, local service providers, which you can find by visiting https://support.apple.com/en-us/HT204039.

 Apple offers unlocked iPhones. Essentially, these phones aren't tied to a particular provider, so you can use them with any of the four iPhone cellular service providers. Although you may save a lot by avoiding a service commitment, purchasing an unlocked phone can be pricey up front.

What's in the Box

When you fork over your hard-earned money for your iPhone, you'll be left holding one box, but that box does include some magical goodies.

Here's what you'll find when you take off the shrink wrap and open the box:

TIP

» **iPhone:** Your iPhone is covered in a thick, plastic-sleeve thingy. Take it off and toss it back in the box.

   Save all the packaging until you're certain you won't return the phone. Apple's standard return period is 14 days.

» **Documentation (and I use the term loosely):** This typically includes a small pamphlet, a sheet of Apple logo stickers, and a few more bits of information.

» **Lightning-to-USB-C cable:** Use this cable to connect the iPhone to your computer (if your computer has a USB-C port) or to a USB-C power adapter (not included).

The iPhone box is a study in Zen-like simplicity. Where's the charging plug? Apple now feels that just about everyone has several chargers laying around their home, so they think it's wasteful (not to mention a little more expensive) to include one with every new iPhone. If you need a charger, Apple will certainly sell you one, and many third-party options are available as well.

TIP

Search for iPhone accessories online. You'll find iPhone covers and cases (from leather to silicone), car chargers, and screen guards to protect your phone's screen.

Take a First Look at the Gadget

In this section, I give you a bit more information about the buttons and other physical features of the newest iPhone models. **Figure 1-3** shows you where each of these items is located on the iPhone 14, 14 Plus, 14 Pro, and 14 Pro Max.

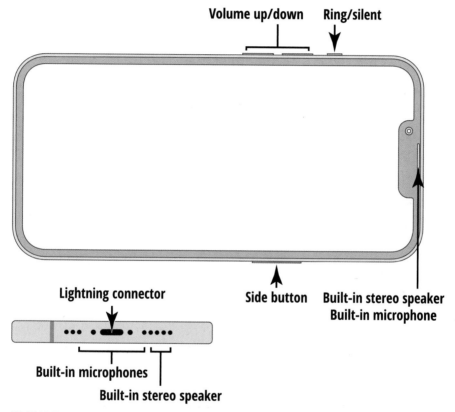

FIGURE 1-3

Here's the rundown on what the various hardware features for iPhones without Home buttons (including all iPhone 14 models) are and what they do.

TIP

If you have an iPhone model with a Home button, read your iPhone's documentation or visit `https://support.apple.com/iphone` to discover the hardware features specific to your device.

» **Side button:** You can use this button to power up your iPhone, put it in sleep mode, wake it up, lock it, force a restart, power it down, and much more.

» **Lightning connector:** Use the Lightning connector to charge your battery (with the Lightning-to-USB-C cable), listen to audio with EarPods (not included), or sync your iPhone with your computer. (See Chapter 4 for more on syncing.)

» **Ring/silent switch:** Slide this little switch to mute or unmute the sound on your iPhone.

» **Built-in stereo speakers:** The speakers in iPhones provide rich stereo sound and deeper bass than previous models, and are located on the bottom edge of the phone and at the top part near the earpiece.

» **Volume up/down buttons:** Tap the volume up button for more volume and the volume down button for less. (You can use the volume up or volume down button as a camera shutter button when the camera is activated.)

» **Built-in microphones:** Built-in microphones make it possible to speak into your iPhone to deliver commands or content. This feature allows you to make phone calls, use video calling services (such as Skype or Zoom), and work with other apps that accept audio input, such as the Siri built-in assistant.

Chapter **2**

Exploring the Home Screen

won't kid you: You're about to encounter a slight learning curve if you're coming from a more basic cellphone. (But if you've owned another smartphone, you have a good head start.) For example, your previous phone might not have had a multi-touch screen and onscreen keyboard.

The good news is that getting anything done on the iPhone is simple, once you know the ropes. In fact, using your fingers to do things is an intuitive way to communicate with your computing device, which is just what the iPhone is.

In this chapter, you turn on your iPhone, register it, and then take your first look at the Home screen. You also practice using the onscreen keyboard, see how to interact with the touchscreen in various ways, get pointers on working with cameras, and get an overview of built-in applications (more commonly referred to as apps).

TIP

Although the iPhone's screen has been treated to repel oils, you're about to deposit a ton of fingerprints on your iPhone — one downside of a touchscreen device. So you'll need to clean the screen from time to time. A soft cloth, like the microfiber cloth you might use to clean eyeglasses, is usually all you'll need to clean things up. Never use harsh chemicals.

What You Need to Use the iPhone

At a minimum, you need to be able to connect to the internet to take advantage of most iPhone features, which you can do using a Wi-Fi network (a network that you set up in your home through an internet service provider or access in a public place such as a library) or a cellular data connection from your cellular provider. You might want to have a computer so that you can connect your iPhone to it to download photos, videos, music, or applications and transfer them to or from your iPhone through a process called *syncing.* (See Chapter 4 for more about syncing.) An Apple service called iCloud syncs content from all your Apple devices (such as the iPhone or iPad), so anything you buy on your iPad that can be run on an iPhone, for example, will automatically be *pushed* (in other words, downloaded and installed) to your iPhone. In addition, you can sync without connecting a cable to a computer by using a wireless Wi-Fi connection to your computer.

Your iPhone will probably arrive registered and activated. If you bought it in a store, the person helping you can usually handle those tasks.

For iPhone SE (second generation or later), 8, 8 Plus, X, XR, XS, XS Max, and all 11, 12, 13, and 14 models, Apple recommends that you have

» A Mac or PC with a USB 2.0 or 3.0 port and one of these operating systems:

- macOS version 10.11.6 (El Capitan) or newer
- Windows 7 or newer

» iTunes 12.8 or newer on a Mac running macOS El Capitan (10.11.6) through macOS Mojave (10.14.6), Finder on Mac's running macOS Catalina (10.15) and newer, and iTunes 12.10.10 or newer on a PC, available at www.itunes.com/download. Windows 10 and 11 users can download iTunes via the Microsoft Store.

» An Apple ID

» Internet access

Turn On iPhone for the First Time

The first time you turn on your iPhone, it will probably have been activated and registered by your cellular carrier or Apple, depending on whom you've bought it from. Follow these steps:

1. **Press and hold down the side button (found a little bit below the top of the upper-right side of newer iPhone models) or the top button (on the first-generation iPhone SE and earlier models) until the Apple logo appears.**

A screen appears, asking you to enter your Apple ID.

2. **Enter your Apple ID.**

If you don't have an Apple ID, follow the instructions to create one.

3. **Follow the series of prompts to set up initial options for your iPhone.**

You can make choices about your language and location, using iCloud (Apple's online sharing service), whether to use a passcode, connecting with a network, and so on.

TIP

You can choose to have personal items transferred to your iPhone from your computer when you sync the two devices using iTunes or Finder, including music, videos, downloaded apps, audiobooks, e-books, podcasts, and browser bookmarks. Contacts and Calendars are downloaded via iCloud, or (if you're moving to iPhone from an Android phone) you can download an Apple app called Move to iOS from the Google Play Store to copy your current Android settings to your iPhone. (Apple provides more information about migrating from Android to iOS at `https://support.apple.com/en-us/HT201196`.) You can also transfer to your computer any content you download directly to your iPhone by using iTunes, the App Store, or non-Apple stores. See Chapters 14 and 16 for more about these features.

Meet the Multi-Touch Screen

When the iPhone Home screen appears, you see a colorful background and two sets of icons, as shown in **Figure 2-1**.

One set of icons appears on the dock, which is along the bottom of the screen. The *dock* contains the Phone, Safari, Messages, and Music app icons by default, though you can swap out one app for another. You can add new apps to populate as many as 14 additional Home screens, for a total of 15 Home screens. The dock appears on every Home screen.

Other icons appear above the dock. (I cover all these icons in Chapter 3.) Different icons appear in this area on each Home screen. You can also nest apps in folders, which gives you the ability to store even more apps on your iPhone, depending on your phone's memory.

TIP

Treat the iPhone screen carefully. It is made of glass and will break if an unreasonable amount of force is applied.

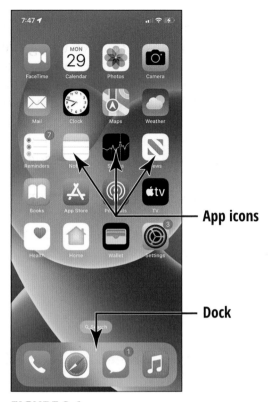

App icons

Dock

FIGURE 2-1

The iPhone uses *touchscreen technology:* When you swipe your finger across the screen or tap it, you're providing input to the device just as you do to a computer by using a mouse or keyboard. You hear more about the touchscreen in the next task, but for now, go ahead and play with it — really, you can't hurt anything. Use the pads of your fingertips (not your fingernails) and try the following:

» **Tap the Settings icon.** The various settings categories appear, as shown in **Figure 2-2**. (You read more about these settings throughout this book.)

REMEMBER

To return to the Home screen, press the Home button or, if you have an iPhone without a Home button, swipe up from the very bottom edge of your screen.

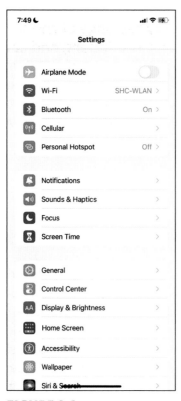

FIGURE 2-2

TIP

» **Swipe a finger from right to left on the Home screen.** This action moves you to the next Home screen.

The little white dots at the bottom of the screen, above the dock icons, indicate which Home screen is displayed. If you see the search field instead, just lightly move your finger on your iPhone's screen and the dots will appear in its place.

» **To experience the screen rotation feature, hold the iPhone firmly while turning it sideways.** The screen flips to the horizontal (or landscape) orientation, if the app you're in supports it.

To flip the screen back, just turn the device so that it's short side is up again (portrait mode). Some apps force iPhone to stay in one orientation or the other.

» **Drag your finger down from the very top edge of the screen to reveal such items as notifications, reminders, and calendar entries.** Drag up from the very bottom edge of the Home screen to hide these items. Then drag up (iPhones with a Home button) or swipe down from the top-right corner to the center (iPhone without a Home button) to display Control Center, which contains commonly used controls and tools.

DISCOVER HAPTIC TOUCH AND QUICK ACTIONS

Haptic touch uses your iPhone's built-in taptic engine to provide haptic (or touch) feedback when you press and hold down on an area or item on your iPhone's screen. For example, if you press and hold down on an icon on the Home screen, a menu of options and tasks will appear, and you'll also feel a tap from your iPhone. Another of my favorite examples of haptic touch is when you press and hold down on the flashlight icon in the lower left of the lock screen. This action causes the flash on the back of your iPhone to turn on or off, and the haptic feedback feels almost like pressing the button on an actual flashlight.

Quick actions involve pressing and holding down on an icon on the screen to see items you're likely to want to select. For example, if you press and hold down on (rather than tap) the Phone icon, you'll get a shortcut list of several call-related options, as shown in the figure. Quick actions provide a shortcut menu to your most frequently used items, saving you time and effort.

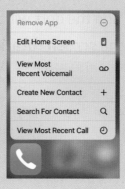

Say Hello to Tap and Swipe

You can use several methods for getting around and getting things done in iPhone by using its multi–touch screen, including

» **Tap once.** To open an application on the Home screen, choose a field (such as a search box), choose an item in a list, use an arrow to move back or forward one screen, or follow an online link, tap the item once with your finger.

» **Tap twice.** Use this method to enlarge or reduce the display of a web page (see Chapter 12 for more on using the Safari web browser) or to zoom in or out in the Maps app.

» **Pinch.** As an alternative to the tap-twice method, you can pinch your fingers together or move them apart on the screen (see **Figure 2-3**) when you're looking at photos, maps, web pages, or email messages to quickly reduce or enlarge them, respectively. This method allows you to grow or contract the screen to a variety of sizes rather than a fixed size, as with the double-tap method.

TIP

Use a three-finger tap to zoom your screen even larger or use multitasking gestures to swipe with four or five fingers. This method is handy if you have vision challenges. Go to Chapter 10 to discover how to turn on this feature using Accessibility settings.

» **Drag to scroll (known as *swiping*).** When you touch your finger to the screen and drag to the right or left, the screen moves (see **Figure 2-4**). Swiping to the left on the Home screen, for example, moves you to the next Home screen. Swiping up while reading an online newspaper moves you down the page; swiping down moves you back up the page.

» **Flick.** To scroll more quickly on a page, quickly flick your finger on the screen in the direction you want to move.

» **Tap the status bar.** To move quickly to the top of a list, a web page, or an email message, tap the status bar at the top of the iPhone screen. (For some sites, you have to tap twice to get this to work.)

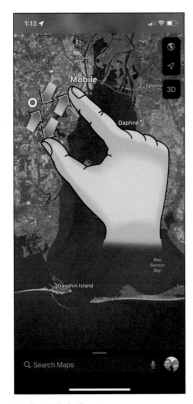
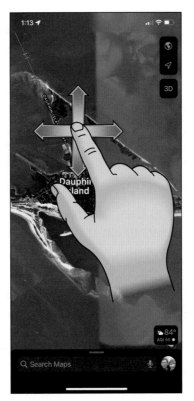

FIGURE 2-3　　　　　**FIGURE 2-4**

> » **Press and hold down.** If you're using Notes or Mail or any other
> application that lets you select text, or if you're on a web page,
> pressing and holding down on text selects a word and displays
> editing tools that you can use to select, cut, or copy and paste
> the text.

When you rock your phone backward or forward, the background moves as well (a feature called *parallax).* You can disable this feature if it makes you seasick. From the Home screen,
tap Settings⇨Accessibility⇨Motion and then turn on the Reduce
Motion setting by tapping the toggle switch (it turns green when
the option is enabled).

Your iPhone enables you to perform *bezel gestures,* which involve sliding left to right from the very outer edge of the phone on the glass to go backward and sliding right to left to go forward in certain apps.

You can try these methods now:

» Tap the Safari icon on the dock at the bottom of any iPhone Home screen to display the Safari web browser.

» Tap a link to move to another page.

» Double-tap the page to enlarge it; then pinch your thumb and finger together on the screen to reduce its size.

» Drag one finger up and down the page to scroll.

» Flick your finger quickly up or down on the page to scroll more quickly.

» Press and hold down your finger on a word that isn't a link. (Links take you to another location on the web.) The word is selected, and the tools shown in **Figure 2-5** are displayed. (You can use this tool to either get a definition of a word or copy it.)

» Press and hold down your finger on a link or an image. A menu appears (shown in **Figure 2-6**) with commands that you select to open the link or picture, open it in a new tab, open it in a tab group, download a linked file, add it to your reading list, copy it, or share it. If you press and hold down on an image, the menu also offers the Add to Photos command. Tap outside the menu to close it without making a selection.

» Position your thumb and finger slightly apart on the screen and then pinch your thumb and finger together to reduce the page. With your thumb and finger already pinched together on the screen, move them apart to enlarge the page.

» Press the Home button or swipe up from the bottom of the screen (iPhone without a Home button) to go back to the Home screen.

FIGURE 2-5

FIGURE 2-6

Browsing the App Library

App Library is an organizational tool that houses every app on your iPhone and organizes them automatically according to categories (defined by Apple). This allows you to hide apps from your Home screens, reducing the number of them you have to scroll through to find an app. You can even hide entire Home screens too! And if you want to display the apps or Home screens again, it's simple to do so. Follow these steps to get around in App Library:

1. Swipe from right to left on any Home screen until the App Library screen appears.

    App Library (shown in **Figure 2-7**) always resides on the screen immediately following the last Home screen.

App Library organizes apps according to categories, such as Suggestions, Recently Added, and Social. Note that each category displays three larger app icons and four smaller app icons, with the exception of Suggestions, which displays four larger icons.

2. Tap one of the larger icons to open the app. When finished, press the Home button (or for iPhones without a Home button, swipe up on the screen) to exit the app and return to the App Library screen.

3. Tap one of the smaller icons to expand the category, and then tap the app you want to open. Tap near the top or bottom of the screen when viewing an expanded category to return to App Library's main screen.

4. Tap the search field at the top of App Library to see all apps listed alphabetically, as shown in **Figure 2-8**. To find a specific app, type its name using the keyboard.

FIGURE 2-7

FIGURE 2-8

5. To perform a quick action on an app icon in App Library, press and hold down on an app icon to open the quick actions menu, the contents of which will vary depending on what options the app offers.

For example, **Figure 2-9** shows the Instagram quick actions menu, which offers some options that aren't available in the Camera app's quick actions menu.

6. To remove an app from your Home screen (leaving it in App Library) or delete it from your iPhone, press and hold down on an app's icon on the Home screen until the quick actions menu appears. Tap Remove App and then tap one of the selections shown in **Figure 2-10**. Tap Delete App to delete the app from your iPhone altogether, or tap Remove from Home Screen to remove the app from the Home screen but retain it in App Library.

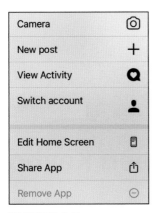

FIGURE 2-9

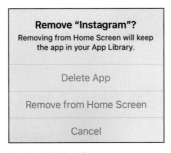

FIGURE 2-10

7. To hide an entire Home screen, press and hold down on any area on a Home screen until all the icons begin to jiggle. Tap one of the small white dots just above the dock to view thumbnails of each of your Home screens, as shown in **Figure 2-11**. Tap the circle below a Home screen to hide or display it. (A visible Home screen displays a check mark in the circle.) Tap Done when you're finished.

The Home screen can be displayed again later when you're ready to return it.

FIGURE 2-11

Display and Use the Onscreen Keyboard

The built-in iPhone keyboard appears whenever you're in a text-entry location, such as a search field or a text message. Follow these steps to display and use the keyboard:

1. Tap the Notes icon on the Home screen to open the Notes app.

2. Open a note you want to work in:

- Tap the New Note icon in the lower-right to create a new note.
- If you've already created some notes, tap one to display the page, and then tap anywhere on the note.

3. Type a few words using the keyboard, as shown in **Figure 2-12**.

FIGURE 2-12

TIP

To make the keyboard display as wide as possible, rotate your iPhone to landscape (horizontal) orientation. (If you've locked the screen orientation in Control Center, you have to unlock the screen to do this.)

TIP

QuickType provides suggestions above the keyboard as you type. You can turn this feature off or on by tapping and holding down on either the emoji icon (smiley face) or the international icon (globe) on the keyboard to display a menu. Tap Keyboard Settings and then toggle the Predictive switch to turn the feature off or on (green). To quickly return to Notes from Keyboard Settings, tap Notes in the upper-left corner of your screen.

Keyboard shortcuts

After you open the keyboard, you're ready to use it for editing text. A number of shortcuts for editing text are available:

» If you make a mistake while using the keyboard — and you will, especially when you first use it — tap the delete key to delete text to the left of the insertion point. (The delete key is near the bottom right, with an *x* on it.)

To type a period and space, just double-tap the spacebar.

» To create a new paragraph, tap the Return key (just like the keyboard on a Mac, or the Enter key on a PC's keyboard).

» To type numbers and symbols, tap the number key (labeled 123) on the left side of the spacebar (refer to Figure 2-12). The characters on the keyboard change (see **Figure 2-13**).

If you type a number and then tap the spacebar, the keyboard returns to the letter keyboard automatically. To return to the letter keyboard tap the ABC key on the left side of the spacebar.

FIGURE 2-13

» Press the Home button or swipe up from the bottom of the screen (iPhone without a Home button) to return to the Home screen.

The shift key

Use the shift key (upward-facing arrow in the lower-left corner of the keyboard) to type capital letters:

» Tapping shift once capitalizes only the next letter you type.

» Double-tap (rapidly tap twice) the shift key to turn on the caps lock feature so that all letters you type are capitalized until you turn the feature off. Tap the shift key once to turn off caps lock.

You can control whether double-tapping the shift key enables caps lock by opening the Settings app, tapping General, tapping Keyboard, and then toggling the switch called Enable Caps Lock.

» To type a variation on a symbol or letter (for example, to see alternative presentations for the letter *A* when you press the A key on the keyboard), hold down on the key; a set of alternative letters or symbols appears (see **Figure 2-14**).

FIGURE 2-14

Emojis

Tap the smiley-faced emoji key to display the emoji keyboard, which contains illustrations that you can insert, including numerical, symbol, and arrow keys, as well as a row of symbol sets along the bottom of the screen. Tapping one of these displays a portfolio of icons from smiley faces and hearts to pumpkins, cats, and more. Tap the ABC key to close the emoji keyboard and return to the letter keyboard.

If you've enabled multilanguage functionality in iPhone Settings, a small globe symbol will appear instead of the emoji key on the keyboard.

QuickPath

QuickPath allows you to zip your finger from key to key to quickly spell words without lifting your finger from the screen. For example, as shown in **Figure 2-15**, spell the word *path* by touching *p* on the keyboard, and then quickly moving to *a* and then *t* and then *h*. Ta-da! You've spelled *path* without your finger leaving the screen.

FIGURE 2-15

Swipe to Search

The Search feature in iOS helps you find suggestions from the web, Music, iTunes, and the App Store as well as suggestions for nearby locations and more. Here's how to use Search:

1. Swipe down on any Home screen (but not from the very top of the screen) or tap the Search button just above the dock to reveal the Search feature (see **Figure 2-16**).

2. Begin entering a search term.

   In the example in **Figure 2-17**, after I typed the word *cammies old,* the Search feature displayed links and other search results. As you continue to type a search term or phrase, the results narrow to match it.

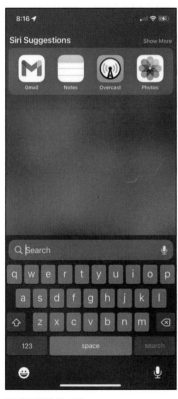

FIGURE 2-16

FIGURE 2-17

3. Scroll down to view more results.

4. Tap an item in the search results to open it in its appropriate app or player.

Wonderful Widgets

Widgets are features that provide snippets of information, such as the weather and calendar appointments, that are provided at a glance so that you don't have to open individual apps. You can add widgets to any Home screen.

To add a widget:

1. Press and hold down on any Home screen until all the icons are jiggling.

2. Tap + in the upper-left corner of the screen.

3. Scroll through the list of widgets and tap the one you want to add, or use the Search Widgets field to find a specific widget.

4. If the widget has several styles (as indicated by dots at the bottom of the screen), like Weather Forecast in **Figure 2-18**, swipe right or left to see those styles. Then tap to select one.

5. Tap the blue Add Widget button.

   You can move your widget by dragging it, or remove it and start over by tapping the – button in the upper left of the widget.

6. When you're finished, tap Done in the upper right.

FIGURE 2-18

Change and Customize Your Lock Screen

The lock screen is the screen you see every time you wake your iPhone. By default, it displays the date, time, other information such as notifications, a flashlight icon, and a camera icon. And that's pretty much been how the lock screen has looked for years. Now Apple has given us the option of making the lock screen look and act how we want it to, so let's see how to do just that!

1. Wake your iPhone by pressing the side button or tapping the screen. The screen you see now is the lock screen.

2. Long-press the screen (press it with your finger for a second or so) to enter lock screen edit mode, shown in **Figure 2-19**.

3. Swipe right or left to view preconfigured lock screens, and tap the one you'd like to use.

FIGURE 2-19

I think we're all agreed that the preconfigured screens are great, but wouldn't you like to add a little personal style? Let me show you how:

1. Long-press the lock screen to enter edit mode.

2. Tap the Customize button below the lock screen you want to use.

3. Tap any of the editable areas that appear on the screen, such as the time or date, or tap the Add Widgets button, as shown in **Figure 2-20**.

4. Make any adjustments you like, then tap the Done button in the upper-right corner to make your changes stick.

FIGURE 2-20

Again, the preconfigured screens are the bee's knees, but I say we try something completely new. To make your own lock screen from scratch:

1. Long-press the lock screen to enter edit mode.

2. Tap the blue circle containing the + in the lower-right corner to add a new lock screen.

3. In the Add New Wallpaper screen that appears, tap a topic at the top of the page or swipe up and down to see the Featured collections, Suggested Photos, Emoji, and other categories, each customizable.

4. Tap your choice, make any editing choices you want, and then tap the Add button in the upper-right corner of the screen to add your new creation to your lock screen collection.

Chapter **3**

Getting Going

N ow it's time to get into even more aspects of using the iPhone and its interface (how you interact with your device).

In this chapter, you look at updating your iOS version (the operating system that your iPhone uses), multitasking, checking out the cameras, discovering the apps that come preinstalled on your iPhone, and more.

Update the Operating System to iOS 16

This book is based on the latest version of the iPhone operating system at the time of this writing: iOS 16. To be sure that you have the latest and greatest features, update your iPhone to the latest iOS now (and do so periodically to receive minor upgrades to iOS 16 or future versions of the iOS). If you've set up an iCloud account on your iPhone, you'll receive an alert and can choose to install the update or not, or you can update manually:

1. Tap Settings.

   Be sure you have Wi-Fi enabled and that your phone is connected to a Wi-Fi network to perform these steps.

2. Tap General.

3. Tap Software Update (see **Figure 3-1**).

   Your iPhone checks to find the latest iOS version and walks you through the updating procedure if an update is available.

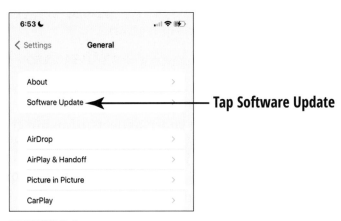

FIGURE 3-1

You can also allow your iPhone to perform automatic updates overnight when one is available. Go to Settings⇨General⇨ Software Update⇨Automatic Updates and toggle the Download iOS Updates, Install iOS Updates, and Security Responses & System Files switches on (green). Your iPhone must be connected to Wi-Fi and its charger to automatically update.

Learn App Switcher Basics

App Switcher lets you easily switch from one app to another without returning to the Home screen. This is accomplished by previewing all open apps and jumping from one to another; you can completely quit an app by simply swiping it upward. To learn the ropes of App Switcher, follow these steps:

1. Open an app.

2. Press the Home button twice. Or for iPhone models without a Home button, drag up from the very bottom of the screen and pause a moment with your finger still on the screen. App Switcher appears and displays a list of open apps (see **Figure 3-2**).

3. To locate another app that you want to switch to, flick to scroll to the left or right.

4. Tap an app to switch to it.

To close App Switcher and return to the app you were working in, press the Home button once. If you have an iPhone without a Home button, tap an app in the list or swipe up from the bottom of the screen to exit App Switcher.

FIGURE 3-2

Examine the iPhone Cameras

iPhones have front- and back-facing cameras. You can use the cameras to take still photos (covered in more detail in Chapter 19) or shoot videos (covered in Chapter 20).

For now, take a quick look at your camera by tapping the Camera app icon on the Home screen. The app opens, as shown in **Figure 3-3**.

You can use the controls on the screen to

» Switch between the front and rear cameras.

» Change from still camera to video camera operation by using the slider at the bottom of the screen.

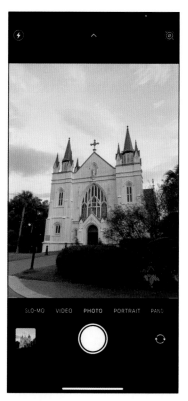

FIGURE 3-3

» Take a picture or start recording a video.

» Choose a 3- or 10-second delay with the timer icon.

» Change aspect ratios.

» Turn HDR (high dynamic range for better contrast) on or off.

» Tap the flash icon to set the flash to On, Off, or Auto.

» Use color filters when taking photos or videos.

» Take a burst of photos by tapping and dragging left or down (depending on the phone's orientation) on the camera's button. A small photo count will display above the button to show how many photos you've taken.

» Open previously captured images or videos.

When you view a photo or video, you can use an iPhone sharing feature to send the image by AirDrop (iPhone 5 and later), Message, Notes, Mail, and other options (depending on which apps you've installed). You can also share through iCloud Photo Sharing, a tweet, Facebook, Instagram, and other apps.

You can do even more things with images: Print them, create a slideshow, use a still photo as wallpaper (that is, as your Home or lock screen background image), assign a still photo to represent a contact. See Chapters 19 and 20 for more detail about using the iPhone cameras.

Take a Look at Face ID

Many newer iPhone models don't have a Home button, so Touch ID (which uses fingerprints to authenticate a user) isn't available. However, they do use a different — and very cool — method of authenticating a user: Face ID. Face ID uses your iPhone's built-in cameras and scanners to scan your face and save a profile of it. It then remembers the information and compares it to whoever is facing the iPhone. If the face doesn't match the profile, the person can't access the iPhone (unless they know and use your passcode, which you have to set up to use Face ID). Face ID is so advanced that it can even work in total darkness.

To set up Face ID:

1. Go to Settings and tap Face ID & Passcode.

2. Tap Set Up Face ID.

3. Hold the iPhone a few inches in front of your face (in portrait mode, not landscape).

4. Tap the Get Started button and then follow the prompts to slowly move your head in a complete circle.

   If you have difficulty moving your head, tap the Accessibility Options button at the bottom of the screen and follow the prompts from there.

5. Tap Continue and follow the prompts to perform the circle step again.

6. Tap Done when finished.

The next time you want to use your iPhone, simply hold it up in front of you, and swipe up from the bottom of the screen when the lock icon unlocks (see **Figure 3-4**). You'll jump right into the Home screen or whatever app you were last using.

FIGURE 3-4

For more information on using Face ID and its capabilities, visit https://support.apple.com/en-us/HT208109.

Discover Control Center

Control Center is a one-stop screen for common features and settings, such as connecting to a network, increasing screen brightness or volume, and even turning the built-in flashlight on or off. Here's how to use it:

1. To display Control Center, swipe up from the very bottom of the screen. If your iPhone doesn't have a Home button, swipe down from the upper-right corner of the screen toward the center.

   The Control Center screen appears, as shown in **Figure 3-5**.

2. In the Control Center screen, tap an icon to access and change a setting, or drag a slider to adjust a setting.

3. To hide Control Center, swipe down from the top of Control Center (iPhone with a Home button) or swipe up from the bottom of the screen (iPhone without a Home button).

Some options in Control Center are hidden from initial view but may be accessed by pressing and holding down on an icon in Control Center. For example, use press and hold down on any of the Communications icons (airplane mode, cellular data, Wi-Fi, and Bluetooth) to reveal two more options: AirDrop and Personal Hotspot (as shown in **Figure 3-6**).

FIGURE 3-5   FIGURE 3-6

Other press-and-hold-down options in Control Center include

>> Adjusting the Flashlight brightness level

>> Selecting a device for AirPlay

>> Setting a quick timer

>> Instantly recording a video or taking a selfie

TIP

Try pressing and holding down on other icons in Control Center to see what other options are waiting for you to discover. If you hold down on an item and its icon pulses just once, no further options are available for the item.

Did you notice the empty space in Control Center when you opened it? That's because you can customize Control Center (a feature I love). All that extra space is waiting to be filled by you:

1. Tap Settings.

2. Tap Control Center to open the Control Center settings, as shown in **Figure 3-7**.

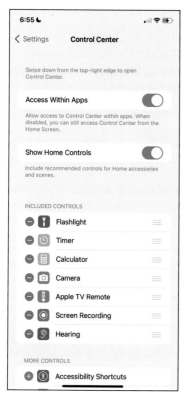

FIGURE 3-7

3. Add or remove items from Control Center:

- To add an item, tap the + in a green circle to the left of the item. You'll see the item in Control Center the next time you visit it.

- To remove an item, tap the − in a red circle to the left of the item, and then tap the Remove button that appears to the right.

Remember to press and hold down on an item to see if the item has any extras. If the item just bounces when you press and hold down, no further options are available.

Lock Screen Rotation

You may have noticed that when you hold your iPhone in portrait mode (narrow and tall) or landscape mode (wide and low), the items on the screen usually flip to match that orientation. However, sometimes you may not want your screen orientation to flip when you move your phone around. Use these steps to lock the iPhone in portrait orientation:

1. Open Control Center.

2. Tap the lock rotation screen icon (padlock with a circular arrow).

   When locked, the icon appears white and the padlock and arrow turn red.

3. To hide Control Center, swipe down from the top of Control Center (iPhone with a Home button) or swipe up from the very bottom of the screen (iPhone without a Home button).

If your iPhone screen doesn't flip when you expect it to, you probably have lock rotation screen enabled. Simply repeat the preceding steps to unlock it.

Explore the Status Bar

Across the top of the iPhone screen is the status bar, as shown in **Figure 3-8**. Tiny icons in this area can provide useful information, such as the time, wireless-connection status, and battery level. Table 3-1 lists some of the most common items on the status bar.

FIGURE 3-8

TABLE 3-1 **Common Status Bar Icons**

Icon	Name	What It Means
(Wi-Fi icon)	Wi-Fi	Your phone is connected to a Wi-Fi network.
(Activity icon)	Activity	A task is in progress — a web page is loading, for example.
2:30 PM	Time	You guessed it: You see the time.
(lock icon)	Screen rotation lock	The screen is locked in portrait orientation and doesn't rotate when you turn the iPhone.
(battery icon)	Battery life	This shows the remaining percentage of battery charge. The indicator changes to a lightning bolt when the battery is charging.

TIP

If you have GPS, cellular, Bluetooth service, or a connection to a virtual private network (VPN), a corresponding symbol appears on the status bar whenever that feature is active. (If you don't know what a virtual private network is, there's no need to worry about it.)

Apple supplies a full list of status bar icons at https://support.apple.com/en-us/HT207354. Keep in mind that icons may not be in the same location on the status bar, depending on whether or not your iPhone has a Home button.

Take Inventory of Preinstalled Apps

The iPhone comes with some applications — or apps, for short — built in. When you look at the Home screen, you see an icon for each app. This task gives you an overview of what each app does.

Several icons appear by default in the dock, which is at the bottom of every Home screen (refer to Figure 3-8). From left to right, these icons are

» **Phone:** Use this app to make and receive phone calls, view a log of recent calls, create a list of favorite contacts, access your voicemail, and view contacts.

» **Safari:** You use the Safari web browser to navigate on the internet, create and save bookmarks of favorite sites, and add web clips to your Home screen so that you can quickly visit favorite sites from there. You may have used this web browser (or another, such as Google Chrome) on your desktop computer.

» **Messages:** If you love to instant message, the Messages app comes to the rescue. You can engage in live text- and image-based conversations with others on their phones or on other devices that use email. You can also send video or audio messages.

» **Music:** Music is the name of your media player. Although its main function is to play music, you can use it also to play audio podcasts and audiobooks.

Apps with icons above the dock on the Home screen include

» **FaceTime:** Have a more personal phone call by using video of the sender and receiver.

» **Calendar:** Use this handy onscreen daybook to set up appointments and send alerts to remind you about them.

» **Photos:** Organize pictures in folders, send photos in email, use a photo as your iPhone wallpaper, and assign pictures to contact records. You can also run slideshows of your photos, open albums, pinch or unpinch to shrink or expand photos, respectively, and scroll photos with a simple swipe.

You can use the photo sharing feature to share photos among your friends. The Photos app displays images by collections, including years and moments.

» **Camera:** As you may have read earlier in this chapter, the Camera app is command central for the still and video cameras built into the iPhone.

» **Mail:** Access email accounts that you've set up in iPhone. Move among the preset mail folders, read and reply to email, and download attached photos to your iPhone. See Chapter 13.

» **Clock:** Display clocks from around the world, set alarms, and use timer and stopwatch features.

» **Maps:** View classic maps or aerial views of addresses and get directions, whether you're traveling by car, by bicycle, on foot, or on public transportation. You can even get your directions read aloud by a spoken narration feature.

» **Reminders:** This useful app centralizes all your calendar entries and alerts. You can also create to-do lists.

» **Notes:** Enter and format text, as well as cut and paste text and objects (such as images) from a website into this simple notepad app.

» **News:** News is a customizable aggregator of stories from your favorite news sources.

» **App Store:** Buy and download apps that do everything from enabling you to play games to building business presentations. Many of these apps and games are free!

» **Podcasts:** Use this app to listen to recorded informational programs. See Chapter 18.

» **TV:** This media player enables you to play videos. The TV app also offers a few features specific to this type of media, such as chapter breakdowns and information about a movie's plot and cast.

- » **Health:** Use this exciting app to record various health and exercise statistics and even send them to your doctor. See Chapter 24 for details.

- » **Wallet:** This Apple Pay feature lets you store a virtual wallet of plane or concert tickets, coupons, credit and debit cards, and more and use them with a swipe of your iPhone across a point of purchase device.

- » **Settings:** Go to this central iPhone location to specify settings for various functions and do administrative tasks, such as setting up email accounts and creating a password.

Some preinstalled apps are located on the second Home screen by default. For example, the following are wrapped up in the Utilities folder: Compass, Calculator, Measure, Magnifier, and Voice Memos. You'll also find the following on the second Home screen:

- » **Files:** This app allows you to browse files stored not only on your iPhone but also on other services, such as iCloud Drive, Google Drive, and Dropbox.

- » **Find My:** The Find My app combines the Find iPhone and Find Friends apps to help you locate Apple devices that you own (see Chapter 25 for more info) and track down friends who also own an Apple device.

- » **Shortcuts:** This new app helps you string together multiple iPhone actions into single commands you can run either manually or by using Siri.

- » **iTunes Store:** Tapping this icon takes you to the iTunes Store, where you can shop 'til you drop (or until your iPhone battery runs out of juice) for music, movies, TV shows, and audiobooks and then download them directly to your iPhone. (See Chapter 16 for more about how the iTunes Store works.)

- » **Weather:** Get the latest weather for your location instantly. You can easily add other locations so you can check the weather where you're going or where you've been.

- » **Stocks:** Keep track of the stock market, including stocks that you personally follow, in real time.

- » **Home:** Control most (if not all) of your home automation devices in one convenient app.

- » **Books:** Because the iPhone has been touted as being a good small-screen e-reader, you should definitely check this one out. (To work with the Books e-reader app itself, check out Chapter 17.)

- » **Translate:** The Translate app translates text or voice between supported languages. See Chapter 9 for more information on how to use this great new tool.

- » **Contacts:** Use this simple app to add, edit, and remove contacts and their information. See Chapter 7 for more details on using the Contacts app.

Several useful apps are free for you to download from the App Store. These include iMovie and iPhoto, as well as the Pages, Keynote, and Numbers apps of the iWork suite.

Put Your iPhone to Sleep or Turn It Off

Earlier in the chapter, I mention how simple it is to turn on the power to your iPhone. Now it's time to put it to sleep (a state in which the screen goes black, though you can quickly wake it up) or turn off the power to give your new device a rest:

- » **Sleep:** Press the side button just below the top of the right side of the phone. The iPhone goes to sleep. The screen goes black and is locked.

The iPhone automatically enters sleep mode after a brief period of inactivity. You can change the time interval at which it sleeps by adjusting the Auto-Lock setting in Settings ⇨ Display & Brightness.

» **Power Off:** From any app or Home screen on an iPhone with a Home button, press and hold down the side button until the Slide to Power Off bar appears at the top of the screen, and then swipe the bar from left to right. You've just turned off your iPhone. If your iPhone doesn't have a Home button, press and hold down both the side button and one of the volume buttons to power off.

» **Force Restart:** If the iPhone becomes unresponsive, you may have to force it to restart.

How you force a restart depends on your iPhone model. To learn more, visit `https://support.apple.com/guide/iphone/force-restart-iphone-iph8903c3ee6/ios`.

To wake most iPhone models up from sleep mode, simply pick up your iPhone (this feature was introduced in iOS 10 and works on the iPhone 6s and newer). Alternatively, on an iPhone model with a Home button, press the Home button once. At the bottom of the screen, the iPhone tells you to press the Home button again. Do so and the iPhone unlocks. If you have an iPhone model without a Home button, simply tap the screen to wake from sleep, or press the side button once.

Chapter **4**

Beyond the Basics

Your first step in having fun with the iPhone is to make sure that its battery is charged. Next, if you want to find free or paid content for your iPhone from Apple, from movies to music to electronic books (e-books) to audiobooks, you'll need to have an iTunes account, also known as an Apple ID.

You can also use the wireless sync feature to exchange content between your computer and iPhone over a wireless network.

Another feature you might take advantage of is the iCloud service from Apple to store and push all kinds of content and data to all your Apple devices — wirelessly. You can pick up where you left off from one device to another through iCloud Drive, an online storage service that enables sharing content among devices so that edits you make to documents in iCloud are reflected in all iPhones and iPads and Macs running OS X Yosemite (version 10.10) or later.

The operating system for Apple's Mac computers used to be called OS X. These days it's referred to as macOS. The Mac operating system is mentioned a few times throughout this book, and you should know that macOS is the new, improved version of OS X.

Charge the Battery by Plugging In

My iPhone showed up in the box fully charged, and let's hope yours did, too. Because all batteries run down eventually, one of your first priorities is to know how to recharge your iPhone battery.

Gather your iPhone and its Lightning–to–USB cable and a power adapter. Then charge your iPhone as follows:

1. Gently plug the Lightning connector end (the smaller of the two connectors) of the Lightning-to-USB cable into the iPhone.

2. Plug the USB end of the Lightning-to-USB cable into the power adapter.

   For iPhones older than the iPhone 12, Apple provided a power adapter in the box.

   If you have an iPhone 12 or later, you can use one of your older power adapters along with a Lightning-to-USB cable. Apple does include a Lightning-to-USB-C cable with the iPhone 12 and newer models. This cable allows for faster charging but requires a USB-C power adapter, which, unfortunately, Apple didn't include. So if you want the fastest charging for an iPhone 12 or a newer model, you need to obtain a USB-C power adapter. (You may have one already if you have an iPhone 11 or a recent iPad or iPad Pro.)

   There's no need to worry if you don't have a power adapter as long as you do have a computer with a compatible USB or USB-C port. You can plug your Lightning-to-USB or Lightning-to-USB-C cable, respectively, into one of those ports to charge.

3. Plug the adapter into an electric outlet.

   The charging icon appears onscreen, indicating that your phone is getting power.

Charge the Battery Wirelessly

Since the iPhone 8, iPhones have the option of *wireless* charging. How does that work? Basically, you set your iPhone on a compatible charging pad and it will charge using induction. Pretty neat. Now the charging pad still has to be plugged in somewhere, so one could technically argue about how wireless this system actually is, but I digress.

So to wirelessly charge your iPhone (model 8 or later), simply set it on a wireless charging pad that's plugged into a wall outlet or USB port of a computer. Some charging pads are a little finicky about phone placement, so wait for the charging symbol to appear onscreen when you set it down. If charging doesn't start, adjust the phone's position.

iPhone 12, 13, and 14 models have a ring of magnets on the back, and these magnets enable you to use Apple's *MagSafe* wireless charger, which looks like a thin puck and snaps onto the back of the iPhone 12, 13, or 14 (and compatible cases). Because MagSafe guarantees perfect placement of the wireless charging pad, the iPhone 12, 13, and 14 can draw more power and charge faster (at 15 watts instead of 7.5 watts for regular wireless charging pads). For more information about MagSafe charging and accessories, go to www.apple.com/shop/accessories/all/magsafe.

Sign into an Apple ID for Music, Movies, and More

The terms *iTunes account* and *Apple ID* are interchangeable: Your Apple ID *is* your iTunes account, but you'll need to be signed in with your Apple ID to download items from the App Store and the iTunes Store.

TIP If you've never set up an Apple ID or an iTunes account, please visit https://support.apple.com/en-us/HT204316 on your computer, iPhone, or iPad for help in doing so.

To be able to download free or paid items from the iTunes Store or the App Store on your iPhone, you must set up an Apple ID. Here's how to sign in to an account after you've created it:

1. Tap Settings on your iPhone.

2. Tap Sign In to Your iPhone, at the top of the Settings list.

3. Enter your Apple ID and password in the fields provided and then tap the Sign In button.

TIP

If you prefer not to leave your credit card info with Apple, one option is to buy an iTunes gift card and provide that as your payment information. You can replenish the card periodically through the Apple Store.

Sign In with Apple is a privacy feature that allows you to use your Apple ID to sign in to any social media account or website. This service provides a simple and secure way to sign in to accounts without having to remember a unique password for each one. Think of it as Apple's more secure replacement for Sign in with Google or Sign in with Facebook, both of which you've probably seen online. The Sign in with Apple button will show up in apps and websites when an account login is required. Check out `https://support.apple.com/en-us/HT210318` for more information on what I consider to be an important and effective privacy feature from Apple.

Sync Wirelessly

You can connect your iPhone to a computer and use the tools there to sync content on your computer to your iPhone. Or, with Wi-Fi turned on in Settings, you can use Wi-Fi to allow cordless syncing if you're within range of a Wi-Fi network that has a computer connected to it.

TECHNICAL
STUFF

If you have a Mac running macOS Mojave (10.14) or earlier, or a Windows-based PC, you'll need iTunes installed on your computer to sync content with your iPhone. You can download iTunes by visiting `www.apple.com/itunes`; Windows 10 and 11 users will be better served downloading iTunes from the Microsoft Store that comes with those versions of Windows. If you have a Mac running macOS Catalina (10.15) or later, you use Finder to sync with your iPhone.

Before you can perform a wireless sync, you need to perform a few steps with your iPhone connected to your computer:

1. If you're charging with an electrical outlet, remove the power adapter.

2. Use the Lightning-to-USB or Lightning-to-USB-C cable to connect your iPhone to your computer.

3. If you have a Mac running macOS Mojave or earlier, or a Windows-based PC running iTunes:

    (a) Open iTunes and click the icon of your iPhone, which appears in the tools in the upper-left corner of the window, as shown in **Figure 4-1**.

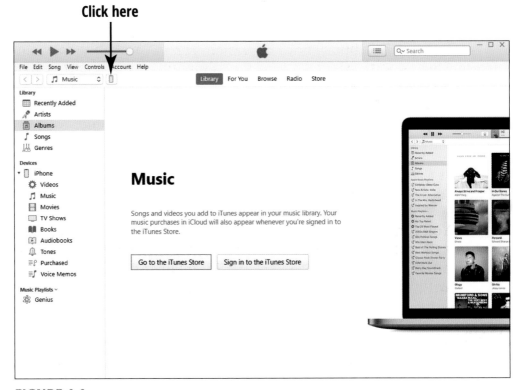

FIGURE 4-1

(b) Select the Sync with This iPhone over Wi-Fi check box, as shown in **Figure 4-2**. You may need to scroll down a bit to see the Sync with This iPhone over Wi-Fi check box.

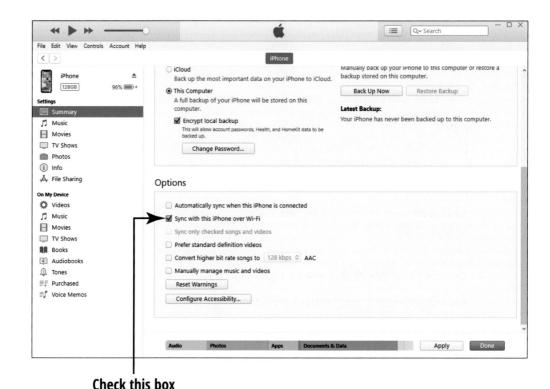

Check this box

FIGURE 4-2

TIP

You can click any item in the Settings section on the left side of the screen shown in Figure 4-2 to handle settings for syncing such items as Movies, Music, and Books. You can also tap the list of items in the On My Device section on the left side to view and even play content directly from your iPhone.

4. If you have a Mac running macOS Catalina or later:

(a) Open a Finder window and then click the icon of your iPhone, which appears in the left sidebar, as shown in **Figure 4-3**.

(b) Select the Show This iPhone When on Wi-Fi check box.

Click here   **Check this box**

FIGURE 4-3

5. Click Apply in the lower-right corner of the iTunes or Finder window.

6. Disconnect your iPhone from your computer.

After you complete the preceding steps, you'll be able to wirelessly sync your iPhone with your computer.

TECHNICAL STUFF

Backing up over a wireless connection is slower than with a cable connected to your computer. If you need to speed things up, simply connect your iPhone to your computer and the sync will continue seamlessly. However, if you then disconnect the iPhone from your computer, the sync will stop. Reconnect to the computer with the cable or with Wi-Fi to pick up where you left off.

TIP

If you have your iPhone set up to sync wirelessly to your Mac or PC, and both are within range of the same Wi-Fi network, your iPhone will appear in your devices list. This setup allows you to sync and manage syncing from iTunes or Finder (depending on which is required for your computer).

Your iPhone will automatically sync once a day if both are on the same Wi-Fi network, iTunes is running (on computers that require iTunes), and your iPhone is charging.

Understand iCloud

There's an alternative to syncing content by using iTunes. Apple's iCloud service allows you to back up most of your content to online storage. That content is then automatically pushed to all your Apple devices wirelessly. All you need to do is get an iCloud account, which is free (again, this is simply using your Apple ID), and choose settings on your devices and in iTunes for which types of content you want pushed to each device. After you do that, content you create or purchase on one device — such as music, apps, and TV shows, as well as documents created in Apple's iWork apps (Pages, Keynote, and Numbers), photos, and so on — is synced among your devices automatically.

WHAT INFORMATION DOES iCLOUD BACK UP?

It's understandable to question what iCloud backs up on your iPhone or other Apple devices, such as your iPad or Mac. Apple's happy to share that with us iPhone fans:

- Device settings
- Home screen settings and app organization
- Photos and videos
- Text messages
- Data from the apps you've installed
- Your purchase history from Apple: apps, music, ringtones, movies, and TV shows
- Apple Watch data
- Your Visual Voicemail password

For more information, please see Apple's support site at https://support.apple.com/en-us/HT207428.

TIP

See Chapter 16 for more about using the Family Sharing feature to share content that you buy online and more with family members through iCloud.

You can stick with iCloud's default storage capacity, or you can increase it if you need more capacity:

» Your iCloud account includes 5GB of free storage. You may be fine with the free 5GB of storage. Content that you purchase from Apple (such as apps, books, music, iTunes Match content, Photo Sharing contents, and TV shows) isn't counted against your storage.

» If you want additional storage, you can buy an upgrade to iCloud+. Currently, 50GB costs only $0.99 per month, 200GB is $2.99 per month, and 2TB (which is an enormous amount of storage) is $9.99 per month. (All prices are in U.S. dollars.) Most likely, 50GB will satisfy the needs of folks who just like to take and share pictures, but if videos are your thing, you may eventually want to consider the larger capacities.

To upgrade your storage, go to Settings, tap your Apple ID at the top of the screen, go to iCloud ⇨ Manage Account Storage, and then tap Change Storage Plan, Buy More Storage, or Upgrade. On the next screen, tap the amount you need and then tap Buy (in the upper-right corner), as shown in **Figure 4-4**.

TIP

If you change your mind, you can get in touch with Apple within 15 days to cancel your upgrade.

FIGURE 4-4

Turn On iCloud Drive

iCloud Drive is the online storage space that comes free with iCloud (as covered in the preceding section).

Before you can use iCloud Drive, you need to be sure that iCloud Drive is turned on. Here's how to turn on iCloud Drive:

1. Tap Settings and then tap your Apple ID at the top of the screen.
2. Tap iCloud to open the iCloud screen.
3. Tap iCloud Drive in the iCloud screen to display the iCloud Drive screen.
4. Tap the Sync This iPhone switch to turn iCloud Drive on (green).

Set Up iCloud Sync Settings

When you have an iCloud account up and running, you have to specify which type of content iCloud should sync with your iPhone. Follow these steps:

1. Tap Settings, tap your Apple ID at the top of the screen, and then tap iCloud.

2. In the Apps Using iCloud section, tap Show All.

3. As shown in **Figure 4-5**, tap the on/off switch for any item that's turned off that you want to turn on (or vice versa).

   You can sync Photos, Mail, Contacts, Calendars, Reminders, Safari, Notes, News, Wallet, Keychain (an app that stores all your passwords and even credit card numbers across all Apple devices), and more. The listing of apps on this screen isn't alphabetical, so scroll down if you don't see what you're looking for at first.

TIP

4. If you want to allow iCloud to provide a service for locating a lost or stolen iPhone, tap iCloud in the top-left corner, tap Apple ID in the top-left corner again, and then tap Find My. Tap Find My iPhone and toggle on the Find My iPhone switch to activate it. While you're there, it's a great idea to toggle on the switches for Find My network and Sent Last Location, as well.

   This service helps you locate, send a message to, or delete content from your iPhone if it falls into other hands. See Chapter 25 for more information.

TIP

Consider turning off the Cellular Data option, which you find in the Cellular section of Settings, to avoid having downloads occur over your cellular connection, which can use up your data allowance. You can turn the option off altogether with the Cellular Data switch (which I don't recommend) or, better, scroll down and disable data for specific apps. Wait until you're on a Wi-Fi connection to have your iPhone perform the downloads.

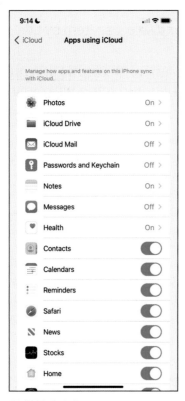

FIGURE 4-5

Browse Your iPhone's Files

Long-time iPhone users have pined for a way to browse files stored on their devices, as opposed to being limited to finding documents and other files only in the apps they're intended for or created by. Finally, a few iOS iterations back, the Files app came along to allow you to browse not only files stored on your iPhone but also stuff you've stored on other online (cloud) services, such as Google Drive, Dropbox, OneDrive, and others.

To familiarize yourself with and use the Files app:

1. Locate the Files app icon and tap to open the app.

2. Tap the Browse button and then do the following on the Browse screen, shown in **Figure 4-6**:

- Tap the search field to search for and select an item by title or content.

- Tap a source in the Locations section and browse a particular service or your iPhone. Once in a source, as shown in **Figure 4-7**, you can tap files to open or preview them, and you can tap folders to open them and view their contents.

- Tap a color in the Tags section to search for and select files you've tagged according to categories.

Browse button

FIGURE 4-6　　　FIGURE 4-7

3. To perform an action on a file (such as duplicating or moving it), tap the more icon (ellipsis), in the upper-right corner of the screen, tap Select, and then tap items to select them for an action. Available actions, found at the bottom of the screen, include duplicating, moving, sharing, and deleting files.

TIP

To retrieve a file you've deleted (you have 30 days to do so before it disappears into the digital ether), go to the Browse screen (tap Browse at the bottom of the screen if you're not already there) and tap Recently Deleted. Tap the more icon (ellipsis) and then tap Select in the upper-right corner. Tap the file you'd like to retrieve, and tap the Recover button at the bottom of the screen. The file will be placed back in the location it was originally deleted from.

SWITCHING FROM ANDROID?

If you're moving from an Android phone to an iPhone, consider downloading the Move to iOS app (developed by Apple) from the Google Play Store on your Android device. This app allows you to wirelessly transfer key content, such as contacts, message history, videos, mail accounts, photos, and more from your old phone to your new one. If you had free apps on your Android device, iPhone will suggest that you download them from the App Store. Any paid apps will be added to your iTunes wish list.

Please read all the information at this page on Apple's support site before using Move to iOS: https://support.apple.com/en-us/HT201196.

IN THIS CHAPTER

» **Monitor how apps are used**

» **Create downtime during your day**

» **Limit how long apps may be used**

» **Restrict access to certain websites**

» **Control privacy settings**

Chapter **5**

Managing and Monitoring iPhone Usage

The iPhone you have will quickly become like an additional limb for you, if it hasn't already. And put one of these things down near a kid and you may likely not see it (or them) for quite a long time. Your iPhone may well become one of the most important tools you own, but your use of it (and that of others) can get out of hand if you're not careful.

Apple understands this concern and has been one of the more active tech companies in helping to resolve this problem. In my opinion, one of the most important features of iOS is Apple's response to the issue of spending too much time on our devices: Screen Time.

The Screen Time feature not only helps you monitor how much time you're spending on your iPhone but also keep track of which apps are consuming your days (and nights). It also can set time limits for app use, lock down your iPhone after certain times, and set content filters to help you or others in your sphere stay away from certain websites or apps.

Meet Screen Time

Screen Time isn't an app unto itself; it's part of the Settings app. To find Screen Time:

1. On the Home screen, tap the Settings icon to open the app.

2. Swipe until you find Screen Time, shown in **Figure 5-1**, and tap to open it. If the Screen Time switch is off (white), tap to turn it on (green), and then proceed through the introduction to enable the feature.

   Once Screen Time is on, you're greeted with a bird's-eye view of your iPhone usage, as shown in **Figure 5-2**. You'll see your iPhone's name as well as a quick glance at total usage time and a graph displaying the length of time you spent using apps of various categories.

   REMEMBER

   You won't see much information in Screen Time if you've just enabled it for the first time or you've had your iPhone for a very short while.

3. To view the Screen Time details screen (see **Figure 5-3**), which presents a more comprehensive listing of your iPhone activities, tap See All Activity, including a breakdown of how you used your apps throughout the day, which apps you used the most, and so on. Next to the Most Used heading, tap Show Categories to see apps listed by category (such as Social, Entertainment, or Productivity & Finance); tap Show Apps & Websites (which replaces Show Categories) to return to the previous view.

Tap here

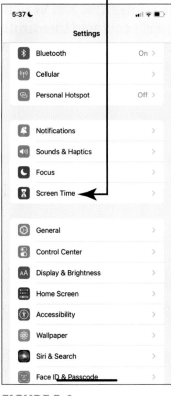

FIGURE 5-1

FIGURE 5-2

TIP

Tap a bar graph for a particular day (designated by S for Sunday, M for Monday, and so forth) to see even more detailed information about an activity for that day. Tap the same bar graph to return to the weekly information breakdown.

TIP

The two tabs at the top of the page are Week and Day. Tap the one you'd like to see displayed.

4. To see app-specific information (the Age Rating may be important to note if you allow children to use your iPhone), tap the app's name in the Most Used section. Tap the back arrow in the upper-left corner to return to the previous screen.

5. Scroll down the page to see how many times you've picked up your iPhone and when (see **Figure 5-4**), get an overview of how many notifications you've received, and which apps generated them.

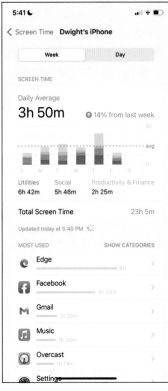

FIGURE 5-3

FIGURE 5-4

TIP

Want to make adjustments based on the activity reported for an app? In the Notifications section, tap an app's name to open the Notifications settings for that app, and make your changes.

6. To return to the main Screen Time settings, tap Screen Time in the upper-left corner.

7. If you'd like to use a passcode to keep your Screen Time settings secure (I recommend it), scroll down to and tap Use Screen Time Passcode and provide a 4-digit passcode.

This passcode will prevent anyone from changing your Screen Time settings and also lets you allow users more time with apps when time limits have been set for them (more on that later in this chapter).

WARNING

Don't forget the Screen Time passcodes you use for Screen Time accounts! Write them down somewhere safe if you need help remembering. If you forget your Screen Time passcode, go to `https://support.apple.com/en-us/HT211021` for help with resetting it. If for some reason you're unable to reset the passcode, your last resort may be to erase your iPhone and set it up as a new device. The moral of the story is try, try, try not to forget this passcode!

8. To view your Screen Time for this device or any other you sign into using your Apple ID, toggle the Share Across Devices switch on.

9. Tap Turn Off Screen Time (at the end of the options) if you want to disable this awesome feature.

Can you tell I very much like this feature of iOS and can't imagine why you'd want to disable it?

Create Some Downtime

Screen Time's Downtime option lets you set aside some time during your day when you use (or should use) your iPhone the least. When Downtime is on, the only apps that will be available are the Phone app and those others you choose to allow. (More on that in the next section.)

iOS 16 allows you to schedule when to use Downtime, and also gives you a quick setting to enable it manually.

To create some iPhone downtime in your day:

1. Open the Screen Time options in the Settings app.

2. Tap Downtime.

3. Determine whether you want to enable Downtime manually or according to a schedule:

- *To enable Downtime manually:* Tap the Turn On Downtime Until Tomorrow button. This leaves Downtime on until midnight. You can disable Downtime manually by tapping the Turn Off Downtime button.

- *To schedule Downtime:* Tap to toggle the Scheduled switch on (green), as shown in **Figure 5-5**. Tap Every Day or Customize Days to decide which days are scheduled for Downtime. Then tap From to select a time to begin your Downtime each day, and tap To to choose a time for Downtime to stop each day. Finally, tap the Turn On Downtime Until Schedule button.

4. To return to the main Screen Time settings, tap Screen Time in the upper-left corner.

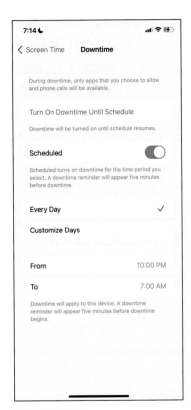

FIGURE 5-5

If you have other iPhones, iPads, or Macs that and are signed into iCloud using the same Apple ID as your iPhone, the Downtime settings will apply for all those devices (if Screen Time is enabled on them).

Allow Certain Apps During Downtime

If you've decided to use Downtime, you may want certain apps to always be available, even during the Downtime period. The Phone app is always available, but you may allow others as you please:

1. Open Screen Time options in the Settings app.

2. Tap the Always Allowed option.

   The Allowed Apps section contains a list of apps that Apple has enabled for use during Downtime (see **Figure 5-6**). These are Phone, Messages, FaceTime, and Maps. Note that Phone can't be disabled.

3. To allow an app to be on during Downtime, scroll through the list in the Choose Apps section and tap the green plus sign (+) to the left of an app's name to add it to the Allowed Apps list.

4. To remove an app from the Allowed Apps list, tap the red minus sign (–) to the left of its name, and then tap the red Remove button that appears to its right.

If you're wondering which apps to allow, consider starting with those used for contacting friends and family as well as those used for medical monitoring and other health-related needs.

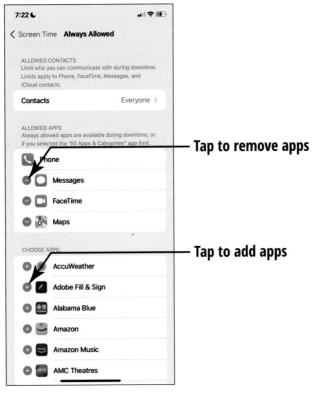

Tap to remove apps

Tap to add apps

FIGURE 5-6

Set App Limits

App Limits is an ingenious feature of Screen Time that helps you curtail excessive use of apps that tend to consume most of your time. Let's face it: Sometimes we get so engrossed in checking out social media and browsing the web that we end up wondering where half the day went. App Limits helps remind you when the time limit is up, but it does allow you to have a bit of extra time or ignore the limit for the day if need be.

To create app limits:

1. Open the Screen Time options in the Settings app.

2. Tap App Limits, and then tap the Add Limit button. If you've previously added App Limits, you'll also see the App Limits switch, which should be toggled on if you want to use the feature.

3. To select all apps in a category, tap the circle to the left of the category. To select a particular app, tap its category to expand it and then tap the circle next to the app. See **Figure 5-7**, where I've expanded the Creativity category.

TIP

To set limits for everything in one fell swoop, tap All Apps & Categories.

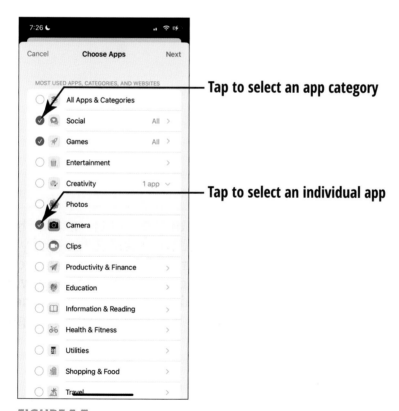

FIGURE 5-7

4. Tap Next in the upper-right corner of the screen.

5. To set time limits in terms of hours and minutes, use the scroll wheel shown in **Figure 5-8**.

To make changes to the allotted time, tap Time and then adjust the scroll wheel to the new time. To set specific times for particular days, tap Customize Days under the scroll wheel. To delete the allotted time, tap the red Delete Limit button at the bottom of the screen, and then tap Delete Limit again to confirm.

6. Tap Add in the upper-right corner and the new limit appears in the App Limits list, as shown in **Figure 5-9**.

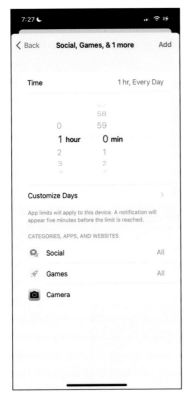

FIGURE 5-8

FIGURE 5-9

7. To add more limits, tap the Add Limit button on the App Limits screen.

8. To exit the App Limits screen, tap the Screen Time button in the upper-left corner.

Your iPhone will notify you when you've reached an app's limit: The screen will become gray, displaying an hourglass (see **Figure 5-10**, left), and the app's icon will be dimmed on its Home screen. If you'd like more time, tap the blue Ignore Limit button at the bottom of the screen, and then tap One More Minute, Remind Me in 15 Minutes, or Ignore Limit for Today (see **Figure 5-10**, right). Tap Cancel if you'd like to adhere to the limit you set for the app.

TIP

If you enabled a passcode for Screen Time, you must enter it before overriding the app's time limit.

FIGURE 5-10

SET COMMUNICATION LIMITS, TOO

Just as you can set limits for how long you can use apps, you can also set limits on who you can communicate with during allowed screen time and/or during downtime. In Settings ⇨ Screen Time, tap Communication Limits. Tap either the During Screen Time or the During Downtime button to customize who you can communicate with during those specific times. During screen time you can elect to communicate with Everyone (default), Contacts Only, or Contacts & Groups with at Least One Contact. During downtime, you can either communicate with Everyone (again, the default) or only specific contacts from your Contacts list. See Chapter 7 for more info on Contacts.

Set Content and Privacy Restrictions

Screen Time helps you prevent access to content that you don't want to be accessed on your iPhone, such as apps, websites, media (movies, music, and so on), and books. You can use it also to set privacy limits.

WARNING

If you're going to restrict content and set privacy restrictions, you should enable the Screen Time passcode — especially if your intention is to protect children who may use your iPhone. This way, only someone who knows the passcode can alter the settings you're about to make.

To set content and privacy options:

1. In the Settings app, open Screen Time options.

2. Tap Content & Privacy Restrictions and then toggle the Content & Privacy Restrictions switch on (green).

3. Tap iTunes & App Store Purchases to allow or block the installation of apps, the deletion of apps, and in-app purchases (purchases that may occur in an app, such as buying upgrades for a character in a game). Also decide whether a password is always required when making

purchases in the iTunes Store or App Store. (I very much recommend using this option if other people use your iPhone.)

4. Tap Back in the upper left to return to the Content & Privacy Restrictions screen, and then tap Allowed Apps.

   This feature allows you to enable or disable apps created by Apple and installed on your iPhone by default. As you can see in **Figure 5-11**, all are enabled to start.

5. If you'd like to disable an app, simply tap the switch to turn it off (white).

   The app is removed from the Home screen; reenabling the app will place it back on the Home screen. This action doesn't delete the app from your iPhone — it only hides it from view while disabled.

6. Tap Back in the upper left to return to the Content & Privacy Restrictions screen, and then tap Content Restrictions.

   The Content Restrictions screen appears.

7. Explore the options in the Content Restrictions screen.

   In the Allowed Store Content section, you can make restrictions based on certain criteria. For example, you can limit which movies are available for purchase or rent in the iTunes Store based on their ratings.

   The Web Content section lets you restrict access to websites. From there, you're able to allow unrestricted access to the web, limit access to adult websites, and further limit access to a list of specific websites that you can customize (as shown in **Figure 5-12)**. You can remove sites from the list by swiping their names to the left and tapping the red Delete button that appears. You can add websites to the list by tapping the blue Add Website button at the bottom of the list.

   Options in the Siri and Game Center sections let you prevent access to untoward content or language, as well as disabling multiplayer games, adding friends to games, turning off the capability to record your iPhone's screen, and more.

Tap to add a site **Tap to delete a site**

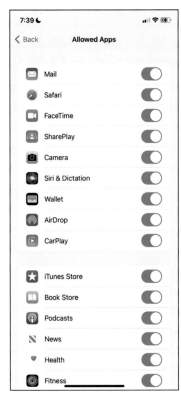

FIGURE 5-11

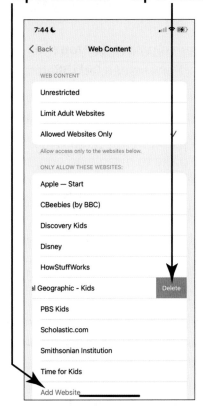

FIGURE 5-12

8. Tap Back in the upper left to return to the Content & Privacy Restrictions screen, and then view the items listed in the Privacy section.

This area lists built-in features and functions. Tapping one shows you which apps are accessing the feature or function, enabling you to limit access to specific apps or to turn off access altogether (by selecting Don't Allow Changes).

The Allow Changes section lets you determine whether changes may be made to such features of your iPhone such as volume limit settings and cellular data options.

9. Tap Back in the upper-left corner to exit the Content & Privacy Restrictions screen.

Manage Children's Accounts

If you have children in your life who use devices linked to your Apple ID through Family Sharing (see Chapter 16), you can manage their activities using Screen Time.

1. In the Settings app, open Screen Time options.

2. Scroll down to the Family section and tap the name of a child's account.

3. Tap Turn On Screen Time and move step by step through the process of enabling Screen Time for the child's account.

4. Tap the Start and End options in the Set Time Away From Screens? page to set times for Downtime, and then tap the Turn On Downtime button or tap Set Up Later to skip.

5. To set app limits for the child's account in the Set App and Website Limits? page, tap the circle next to an app category, or just tap the circle next to All Apps & Categories. To set a time limit, tap Set next to Time Amount, set the desired amount of time, and then tap the Set App Limit button at the bottom of the screen.

   Of course, you can also tap Set Up Later to skip.

6. Set a Screen Time passcode to prevent changes from being made to the settings for this account, if prompted.

   Screen Time is now activated for the child's account. You may make changes to the account's Screen Time settings at any time.

Using Your iPhone

2

IN THIS PART . . .

Calling friends and family

Managing contacts

Messaging and video calls

Discovering utilities

Customizing accessibility

Getting to know Siri

IN THIS CHAPTER

» **Place and end calls**

» **Place calls using Contacts**

» **Receive and return calls**

» **Use Favorites**

» **Use tools during calls**

» **Enable Do Not Disturb**

» **Reply to calls via text and set reminders to call back**

Chapter **6**

Making and Receiving Calls

f you're someone who wants a cellphone only to make and receive calls, you probably didn't buy an iPhone. Still, making and receiving calls is one of the primary functions of any phone, smart or otherwise.

In this chapter, you discover all the basics of placing calls, receiving calls, and using available tools during a call to mute sound, turn on the speakerphone, and more. You also explore features that help you manage how to respond to a call that you can't take at the moment, how to receive calls when in your car, and how to change your ringtone.

Place a Call by Using the Keypad

Dialing a call with a keypad is an obvious first skill for you to acquire, and it's super-simple.

TIP

CarPlay is a feature that provides the capability to interact with your car and place calls by using Siri voice commands and your iPhone. Apps designed for CarPlay allow you to interact with controls on your car's console while keeping your attention on the road. Visit www.apple.com/ios/carplay/available-models to see which auto manufacturers are currently utilizing CarPlay in their products. To date, over 600 models support CarPlay.

To manually dial a call, follow these steps:

1. On any Home screen, tap Phone in the dock, and the app opens. Tap the Keypad icon at the bottom of the screen, and the keypad appears (see **Figure 6-1**).

TIP

If anything other than the keypad appears, you can just tap the Keypad icon at the bottom of the screen to display the keypad.

2. Enter the number you want to call by tapping the number buttons; as you do, the number appears above the keypad. If the person is already in your contacts, that person's name will appear under the number.

TIP

When you enter a phone number, before you place the call, you can tap Add Number (in blue text found directly below the phone number) to add the person you're calling to your Contacts app. You can create a new contact or add the phone number to an existing contact using this feature.

3. If you enter a number incorrectly, you can clear the numbers one at a time by tapping the delete icon (left-pointing arrow with an X).

The delete icon appears on the keypad when you begin entering a number.

4. Tap the call icon (green circle with a white telephone headset).

The call is placed and tools appear, as shown in **Figure 6-2**.

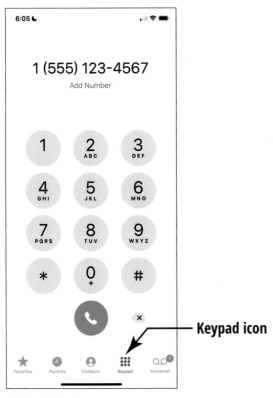

Keypad icon

FIGURE 6-1

FIGURE 6-2

TIP

If you're on a call that requires you to enter numbers or symbols (such as a pound sign), tap the Keypad icon on the tools that appear during a call to display the keypad. See more about using calling tools in the "Use Tools during a Call" task, later in this chapter.

End a Call

In the following tasks, I show you several other ways to place calls; however, I don't want to leave you on your first call without a way out. When you're on a phone call, the green call button (refer to Figure 6-2) changes to a red end call button. Tap the end call button, and the call is disconnected.

Place a Call by Using Contacts

If you've created a contact record that includes a phone number (see Chapter 7), you can use the Contacts app to place a call.

1. Tap the Phone icon in any Home screen dock.

2. Tap the Contacts icon at the bottom of the screen.

3. In the Contacts list that appears, scroll up or down to locate the contact you need, tap a letter along the right side to jump to that section of the list, or press and hold down on the right side and then scroll up or down to the letter you need. You can also search for a contact by tapping the search field and entering part of the contact's name, as shown in **Figure 6-3**.

4. Tap the contact to display their record. In the record that appears (see **Figure 6-4**), tap the phone number field to place the call.

TIP

If you locate a contact and the record doesn't include a phone number, you can add it at this point by tapping Edit in the upper-right corner, entering the number, and then tapping Done (upper-right corner). Place your call following Step 4 in the preceding steps.

Search field

Tap the phone number to call

FIGURE 6-3

FIGURE 6-4

Return a Recent Call

If you want to dial a number from a call you've recently made or received, you can use the Recents call list.

1. Tap Phone in the dock on any Home screen.

2. Tap the Recents icon at the bottom of the screen. A list of recent calls that you've both made and received appears (see **Figure 6-5**).

   Missed calls appear in red.

3. If you want to view only the calls you've missed, tap the Missed tab at the top of the screen.

REMEMBER

4. Tap the info icon (*i* in a small circle) to the right of any item to view information about calls to or from this person or establishment (see **Figure 6-6**). The information displayed here might differ depending on the other phone and connection.

FIGURE 6-5

FIGURE 6-6

5. Tap the Recents icon in the upper-left corner to return to the Recents list, and then tap any call record to place a call to that number.

Use Favorites

You can save up to 50 contacts to Favorites in the Phone app so that you can quickly make calls to your A-list folks or businesses.

1. Tap Phone on any Home screen.

2. Tap the Favorites icon at the bottom of the screen.

3. In the Favorites screen that appears (see **Figure 6-7**), tap + (add) in the upper-left corner.

   Your Contacts list appears, displaying contact records that contain a phone number in bold.

4. Tap a contact that you want to make a favorite. Next, in the menu that appears, tap a mode of communication, depending on which type of call you prefer to make to this person most of the time (see **Figure 6-8**). Tap the arrows to the right of the mode of communication to choose other options for the contact, such as alternate phone numbers.

   The Favorites list reappears with your new favorite contact on it.

FIGURE 6-7

FIGURE 6-8

5. To place a call to a favorite, open the Phone app, tap Favorites, and then tap a person in the list to place a call.

If you decide to place a FaceTime call to a favorite that you've created using the Call setting, just tap the info icon (circled *i*) that appears to the right of the favorite's listing and then tap the FaceTime call button in the contact record that appears. You can also create two contacts for the same person, one with a cellphone and one with a land-line phone, for instance, and place one or both in Favorites. See Chapter 8 for more about making FaceTime calls.

Receive a Call

There's one step to receiving a call. When a call comes in to you, the screen shown in **Figure 6-9** appears:

» Tap Accept to pick up the call.

» Tap Message to send a preset text message (such as "Can I call you later?") without picking up.

Edit the preset messages in Settings under Phone➪Respond with Text.

» Tap Remind Me to have iPhone send you a reminder to return the call without answering it.

» Tap Decline or don't tap any buttons after a few rings, and the call goes to your voicemail.

Incoming calls may also appear in another way. If you're actively using your iPhone and it's unlocked, the call will appear as a banner at the top of the screen, as shown in **Figure 6-10**. Simply tap the green or red button to answer or not answer the call, respectively. You can also swipe down on the incoming call banner to see more call tools, or swipe up on the banner to send it away — the call will go to voicemail after a few rings.

Incoming call banner

FIGURE 6-9

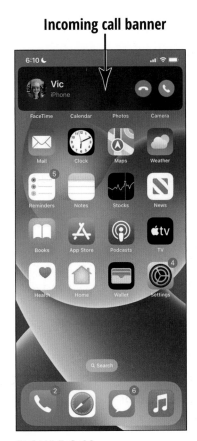

FIGURE 6-10

TIP

There's another quick way to send a call to voicemail or decline it: Press the sleep/wake button on the top or side of your phone once to send a call to voicemail or press the same button twice to decline the call. You can also press one of the volume buttons twice to perform the same task, which might be more convenient.

With the Handoff feature, if your iPhone is near an iPad or a Mac computer (2012 model or later) and a call comes in, you can connect to the call via Bluetooth from any of the three devices by clicking or swiping the call notification. All devices have to be signed into the same iCloud account and have enabled Bluetooth and Handoff (General ⇨ AirPlay & Handoff) under Settings.

BLOCKING AND SILENCING UNSOLICITED CALLERS

Are you receiving calls from people and companies that you'd rather not get? Are telemarketers constantly bothering you? Never fear, Apple is here! You can easily block unwanted or unsolicited callers. Open the Phone app, tap the Recents icon at the bottom of the screen, and then tap the info icon (circled *i*) to the right of the caller you want to block. Swipe to the bottom of the screen, tap the red Block This Caller button, and then tap Block Contact to confirm. Simply follow the same steps to unblock the caller if you'd like to resume reception of their calls. iOS 16 has another feature that goes one step further. If you're getting inundated with tele-marking calls, your iPhone can *silence* unknown calls automatically and send them directly to voicemail. In other words, if you get a call from someone who isn't listed in your contacts, the call will go directly to voicemail, and your iPhone won't bother you with a ring or a vibration. You'll simply see a notification that an incoming call was received, and the caller will appear in your list of recent calls. If the caller leaves a voicemail, you'll receive a notification for that as well.

To turn on the silencing feature, open the Settings app, scroll down to Phone, tap Silence Unknown Callers, and then turn on the Silence Unknown Callers option. For many people, this feature is a godsend. Just make sure that all the important people in your life — including not just family and friends but also coworkers, doctors, insurance agents, pharmacists, school principals, and so forth — are in your contacts. That way, they won't be silenced when they try to reach you. If a silenced caller is someone you want to hear from in the future, add that person's number to your contacts, and your phone will ring or vibrate (or both) when the person calls.

Use Tools During a call

When you're on a call, whether you initiated it or received it, the set of tools shown in **Figure 6-11** appear.

Here's what these six buttons allow you to do, starting with the top-left corner:

>> **Mute:** Silences your microphone so that the caller can't hear you, though you can still hear the caller. The Mute button background turns white, as shown in **Figure 6-12**, when a call is muted. Tap again to unmute the call. You can also tap and hold down on the Mute button to place the call on hold; simply tap the newly visible Hold button to resume.

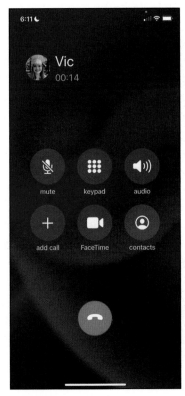

FIGURE 6-11   **FIGURE 6-12**

» **Keypad:** Displays the numeric keypad.

» **Speaker/Audio:** Turns the speakerphone feature on and off, or allows you to change the method used for listening to your call if you're connected to multiple output devices. If you're near a Bluetooth audio device and have Bluetooth turned on in your iPhone Settings, you'll see a list of sources (such as a car Bluetooth connection for hands-free calls) and you can choose the one you want to use.

» **Add Call:** Displays Contacts so that you can add a caller to a conference call.

» **FaceTime:** Begins a video call with someone who has an iPhone 4 or more recent model, iPod touch (fourth generation or later), iPad 2 or third generation or later, an iPad mini, or a Mac running macOS 10.7 or later.

» **Contacts:** Displays a list of contacts.

TIP

You can pair your Apple Watch with your iPhone to make and receive calls using it. For more information, visit Apple's Support site at https://support.apple.com/watch.

Turn On Do Not Disturb

The Do Not Disturb feature has been around for quite a while, but it's now a part of Focus, a feature that keeps you from being disturbed by incoming calls and notifications during various times and tasks. You can customize a list of people or apps that can still contact or notify you, even when a focus is enabled.

Do Not Disturb causes your iPhone to silence incoming calls and notifications when it's locked, displaying only a moon–shaped icon to let you know that a call is coming in. To enable Do Not Disturb:

1. Open Control Center.

If your iPhone has a Home button, swipe up from the bottom of the screen. Otherwise, swipe down from the upper-right corner to the center. The Control Center screen appears.

2. Tap the text of the Focus button, not the icon to the left of the button.

3. To enable the Do Not Disturb feature, tap its button (shown in **Figure 6-13**).

FIGURE 6-13

When you turn on the Do Not Disturb feature, calls from favorites are automatically allowed through by default.

TIP

Set Up a Driving Focus

You can set up a specific focus for driving, which will cause your iPhone to use Do Not Disturb when you enable it manually or if you allow your iPhone to enable it automatically.

Here's how to create a focus for driving:

1. Tap Settings ➪ Focus.

2. Tap + in the upper-right corner.

3. In the What Do You Want to Focus On? screen, tap Driving (see **Figure 6-14**).

FIGURE 6-14

4. Tap the Customize Focus button.

5. To customize how the Driving focus is activated, tap the While Driving button in the Turn On Automatically section near the bottom of the screen and choose one of the following:

   - *Manually (default):* You activate the feature through Control Center (see Chapter 3 for more information).

- *When Connected to Car Bluetooth:* The feature activates automatically when your iPhone connects to your vehicle's Bluetooth.
- *Automatically:* The feature is enabled whenever your iPhone detects that you're in an accelerating vehicle. Yes, it's that smart! This feature could be inconvenient when you're simply a passenger — simply disable the feature for the duration of the ride.

6. Tap Driving Focus in the upper-left corner of the screen, and then tap Focus in the upper-left corner to exit the newly created Driving focus. It's now ready to use!

When you're connected to your vehicle's Bluetooth or a hands-free accessory, phone calls can come through but not texts or notifications.

If your car is equipped with CarPlay, you can have your iPhone automatically activate the Driving focus when you connect it to your car via Bluetooth or USB. Go to Settings ⇨ Focus ⇨ Driving ⇨ While Driving, and toggle on (green) the Activate with CarPlay switch.

Set Up Exceptions for Do Not Disturb

If there are people or apps whose calls or notifications you want to receive even when the Do Not Disturb feature or a specific focus is turned on, you can set up Allowed Notifications:

1. Go to Settings ⇨ Focus.

2. Tap either Do Not Disturb or the particular focus you'd like to edit.

3. To customize who or what app can contact or notify you, even when Do Not Disturb or the particular focus is on, tap the People button or the Apps button or both in the Allow Notifications section, as shown in **Figure 6-15**.

FIGURE 6-15

4. Tap the Allow Notifications From button, and then tap the Add People or Add Apps button (depending on which you selected in the preceding step), as shown in **Figure 6-16**.

5. Tap the names of people or apps for which you want to allow exceptions, and then tap the Done button in the upper-right corner.

   The area below Allow Notifications From displays the list of people or apps for which you've allowed exceptions (see **Figure 6-17**).

FIGURE 6-16   **FIGURE 6-17**

6. To remove an individual person or app from the list, tap – in the upper-left corner of an icon.

7. To exit, tap the button in the upper-left corner.

Reply to a Call via Text or Set a Reminder to Call Back

You can reply with a text message to callers whose calls you can't answer. You can also set up a reminder to call the person back later:

» When a call comes in that you want to send a preset message to, tap Message. Then tap a preset reply, or tap Custom and enter a message.

» When a call comes in for which you want to set up a reminder, tap Remind Me, and then tap In One Hour or When I Leave.

Change Your Ringtone

The music or sound that plays when a call is coming in is called a ringtone. Your phone is set up with a default ringtone, but you can choose among a large number of Apple-provided choices.

1. Tap the Settings icon.

2. Tap Sounds & Haptics and then tap Ringtone.

3. Scroll down the list of ringtones and tap on one to preview it (see **Figure 6-18**).

4. When you've selected the ringtone that you want, tap Back to return to the Sounds & Haptics settings with your new ringtone in effect.

FIGURE 6-18

TIP

You can set custom ringtones also for individual contacts by using the Contacts app. So when, say, your child or children call, you can have a ringtone that you associate only with them. And you can do the same for your spouse, boss, or best friend. Open a contact's record, tap Edit, and then tap the Ringtone setting to display a list of ringtones. Tap one and then tap Done to save it.

IN THIS CHAPTER

» **Add, share, and delete contacts**

» **Sync contacts with iCloud**

» **Assign a photo to a contact**

» **Add social media information**

» **Customize contacts' profiles**

Chapter **7**

Organizing Contacts

C ontacts is the iPhone equivalent of the dog-eared address book or Rolodex that used to sit by your phone. The Contacts app is simple to set up and use yet has some powerful features beyond simply storing names, addresses, and phone numbers.

For example, you can pinpoint a contact's address in iPhone's Maps app. You can use your contacts to address email and Facebook messages and Twitter tweets quickly. If you store a contact record that includes a website, you can use a link in Contacts to view that website instantly. In addition, of course, you can easily search for a contact by a variety of criteria, including how people are related to you, such as family or mutual friends, or by lists you create.

Onscreen context is a feature that allows you to make requests of Siri while viewing a specific contact. Siri will then carry out the request for the contact you're viewing, without the need to reference the person.

In this chapter, you discover the various features of Contacts, including how to save yourself spending time entering contact information by syncing contacts with such services as iCloud.

Add a Contact

1. Access Contacts by tapping Phone at the bottom of the Home screen, and then tap the Contacts icon at the bottom of the screen.

Or simply find and tap the Contacts app icon on a Home screen (it's on the second one by default). An alphabetical list of contacts appears, like the one shown in **Figure 7-1**.

2. Tap + (add) in the upper-right corner.

A blank New Contact page opens (see **Figure 7-2**).

FIGURE 7-1

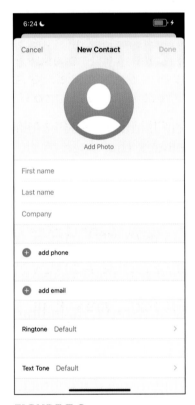

FIGURE 7-2

3. Tap in any field and the onscreen keyboard appears. Enter any contact information you want.

TIP

Only one of the First name, Last name, and Company fields is required, but feel free to add as much information as you like.

4. To scroll down the contact's page and see more fields, flick up on the screen with your finger.

5. If you want to add information (such as a mailing or street address), tap the relevant add field, which opens additional entry fields.

6. To add an information field, such as Nickname or Suffix, tap the blue add field button toward the bottom of the page. In the Add Field dialog that appears (see **Figure 7-3**), choose a field to add, and then populate it with the relevant info.

TIP

To view all the fields, flick up or down the screen.

TIP

If your contact has a name that's difficult for you to pronounce, consider adding the Phonetic First Name, Phonetic Middle Name, or Phonetic Last Name field (or some combination thereof) to the person's record so you can spell out the name phonetically.

7. Tap the Done button in the upper-right corner when you've finish making entries.

The new contact appears in your address book.

8. Tap your new contact to view the details (see **Figure 7-4**).

TIP

You can choose a distinct ringtone or text tone for a new contact. Just tap the Ringtone or Text Tone field in the New Contact form (refer to Figure 7-2) or when editing a contact to see a list of options. Then when the contact calls on the phone, calls using FaceTime, or texts, you will quickly and easily recognize the person from the tone that plays.

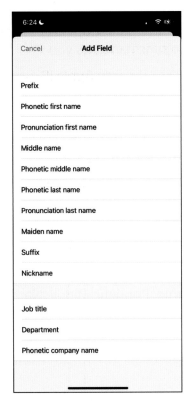

FIGURE 7-3

FIGURE 7-4

Sync Contacts with iCloud

You can use your iCloud account to sync contacts from your iPhone to iCloud to back them up. These also become available to your email account, if you set one up.

TIP

Mac users can also use iTunes or Finder (if your Mac is running macOS Catalina or newer) to sync contacts among all your Apple devices. Windows PC users also use iTunes. See Chapter 4 for more about iTunes settings.

1. On the Home screen, tap Settings, tap the name of your Apple ID account (at the top of the screen), and then tap iCloud.

2. In the iCloud settings screen, tap Show All under the Apps Using iCloud section.

3. In the Apps Using iCloud screen shown in **Figure 7-5**, make sure that the switch for Contacts is on (green) to sync contacts.

4. In the top-left corner of the screen, tap iCloud, tap Apple ID, and then tap the Settings button to return to Settings.

5. To choose which email account to sync with (if you have more than one account set up), scroll down and tap Mail. Tap the Accounts section, and then tap the email account you want to use.

6. In the following screen (see **Figure 7-6**), toggle the Contacts switch on to merge contacts from that account via iCloud.

Tap here

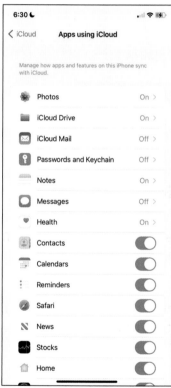

FIGURE 7-5

FIGURE 7-6

Assign a Photo to a Contact

Assigning photos to a contact can be both functional and aesthetic. Here's how to go about doing so:

1. With Contacts open, tap a contact to whose record you want to add a photo.

2. Tap Edit.

3. On the info page that appears (see **Figure 7-7**), tap Add Photo (or Edit if the contact already has a photo assigned).

4. In the menu that appears, tap a suggested photo, tap a memoji, or tap the all photos icon (labeled in **Figure 7-8**) and locate and tap a photo to select it.

Tap to add or edit a photo   Camera   All photos

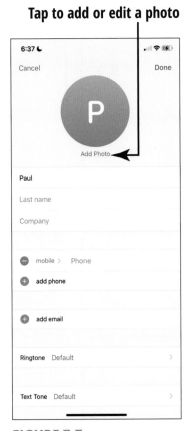

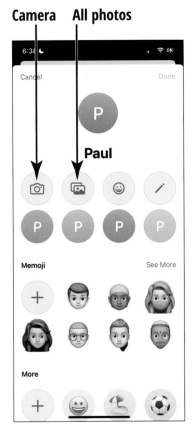

FIGURE 7-7   FIGURE 7-8

TIP

You could also tap the camera icon (labeled in Figure 7-8) to take that contact's photo on the spot.

5. In the Move and Scale dialog that appears (see **Figure 7-9**), position the photo by dragging it with your finger.

TIP

While in the Move and Scale dialog, you can modify the photo before saving it to the contact's information. Pinch or unpinch your fingers on the iPhone screen to contract or expand the photo, respectively, and drag the photo around the space to focus on a particular section.

6. Tap the Choose button to use the photo for this contact. If prompted, you may also select a filter to use with the photo.

7. Tap Done, tap Done again, and finally tap Done once more to save changes to the contact.

The photo appears on the contact's info page (see **Figure 7-10**).

FIGURE 7-9

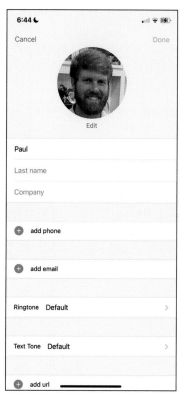

FIGURE 7-10

Add Social Media Information

Your iPhone lets you add social media information to your contacts so that you can quickly tweet (send a short message to) others using Twitter, comment to a contact on Facebook, and more. You can use any of the following social media platforms with Contacts: Twitter, Facebook, Flickr, LinkedIn, Myspace, and Sina Weibo.

To add social media information to contacts, follow these steps:

1. Open Contacts in the Phone app, and then tap a contact.

2. Tap Edit in the upper-right corner of the screen.

3. Scroll down and tap Add Social Profile.

   Twitter is the default service that pops up.

4. To change the default service, tap Twitter and select from the list of services shown in **Figure 7-11**. Tap Done after you've selected the service you'd like to use.

TIP

Other social media and app platforms may also appear in the services list, depending on which apps you have installed on your iPhone.

5. Enter the information for the social profile as needed.

   You may add multiple social profiles if you like.

TIP

6. Tap Done and the information is saved.

   The social profile account is now displayed when you select the contact, and you can send tweets, Facebook messages, or what-have-you by simply tapping the username, tapping the service you want to use to contact the person, and then tapping the appropriate command (such as Facebook posting).

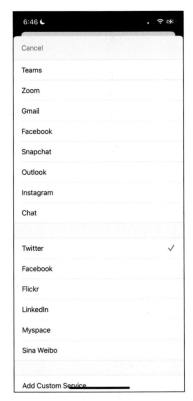

FIGURE 7-11

Designate Related People

You can quickly designate relationships in a contact record if those people are saved to Contacts. One great use for this feature is to then use Siri to, for example, simply say, "Call nephew" to call someone who is designated in your contact information as your nephew.

TIP

There's a setting for Linked Contacts in the Contacts app when you're editing a contact's record. Using this setting isn't like adding a relation; rather, if you have records for the same person that have been imported into Contacts from different sources, such as Google or Twitter, you can link them to show only a single contact.

1. Tap a contact and then tap Edit.

2. Scroll down the record and tap Add Related Name.

   The field labeled Mother (see **Figure 7-12**) appears.

3. If the contact you're looking for is indeed your mother, leave it as is. Otherwise, tap Mother and select the correct relationship from the list provided.

4. Tap the blue info icon to the right of the Related Name field, and your Contacts list appears. Tap the related person's name, and it appears in the field (see **Figure 7-13**).

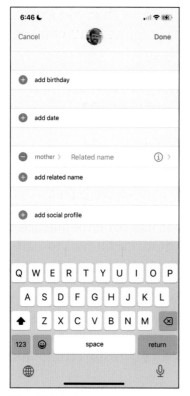

FIGURE 7-12

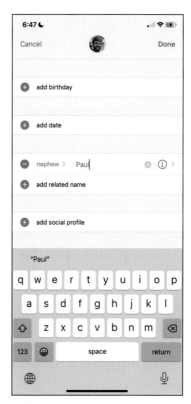

FIGURE 7-13

5. If you would like to add additional names (such as a nickname), tap Add Related Name again and continue to add additional names as needed.

6. Tap Done to complete the edits.

After you add relations to a contact record, when you select the person in the Contacts main screen, all the related people for that contact are listed there.

Set Individual Ringtones and Text Tones

If you want to hear a unique tone when you receive a phone or Face-Time call from a particular contact, you can set up this feature in Contacts. For example, if you want to be sure that you know instantly whether your spouse, friend, or boss is calling, you can set a unique tone for that person.

If you set a custom tone for someone, that tone will be used when that person calls or contacts you via FaceTime too.

To set up custom tones, follow these steps:

1. Tap + in the upper-right corner of the Contacts app to add a new contact, or select a contact in the list of contacts and tap Edit.

2. Tap the Ringtone field.

A list of tones appears (see **Figure 7-14**).

You can set a custom text tone to be used when the person sends you a text message. Tap Text Tone instead of Ringtone in Step 2, and then follow the remaining steps.

3. Scroll up and down to see the full list of tones. Tap a tone, and a preview of the tone plays. When you hear one you like, tap Done in the upper-right corner.

FIGURE 7-14

TIP

If your Apple devices are synced via iCloud, setting a unique ringtone for an iPhone contact also sets it for use with FaceTime and Messages on your iPad, Mac, and Apple Watch. See Chapter 4 for more about iCloud.

Search for a Contact

If you're like a lot of folks, your contacts list can grow exponentially in a short time. Being able to search for a contact, as opposed to scrolling through them all, is a real boon.

1. With Contacts open, tap the search field at the top of your Contacts list (see **Figure 7-15**).

The onscreen keyboard opens.

2. Type the first letters of either the first or last name, the nickname, the phonetic name, or the company.

All matching results appear, as shown in **Figure 7-16**. For example, typing *App* might display Johnny Appleseed and Apple Inc. in the results, both of which have *App* as the first three letters of the first or last part of the name or address.

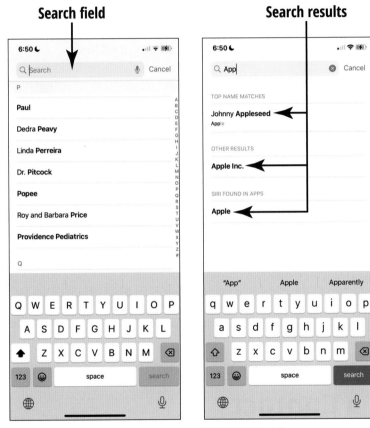

FIGURE 7-15   FIGURE 7-16

TIP

You can use the alphabetical listing along the right side of All Contacts and tap a letter to locate a contact. Also, you can tap and drag to scroll down the list of contacts on the All Contacts page.

3. Tap a contact in the results to display that person's info page.

TIP

You can search by phone number simply by entering the phone number in the search field until the list narrows to the person you're looking for. This might be a good way to search for all contacts in your town or company, for example.

Share a Contact

After you've entered contact information, you can share it with others via an email, a text message, and other methods.

1. With Contacts open, tap a contact name to display its information.

2. On the information page, scroll down and tap Share Contact. In the dialog that appears, shown in **Figure 7-17**, tap the method you'd like to use to share the contact.

TIP

If you want to send only some of the contact's information, tap the Filter Fields button and select only the information you desire to share.

3. To share with a nearby AirDrop-enabled device, use the AirDrop button on the screen shown in Figure 7-17. Just select a nearby device, and your contact is transmitted to that person's Apple device (an iPhone, a Mac with the AirDrop folder open in Finder, or an iPad).

4. If emailing or sharing via text message, use the onscreen keyboard to enter the recipient's information.

5. If sharing an email or a text message, tap the Send button.

The message goes to your recipient with the contact information attached as a .vcf (vCard) file, as shown in **Figure 7-18**. (The vCard format is commonly used to transmit contact information.)

TIP

When someone receives a vCard containing contact information, that person needs only click the attached file to open it. At this point, depending on the email or contact management program, the recipient can perform various actions to save the content. Other iPhone, iPod touch, iPad, or Mac users can easily import .vcf records as new contacts in their own Contacts apps.

Attached vCard

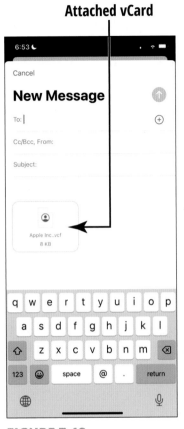

FIGURE 7-17   FIGURE 7-18

Delete a Contact

When it's time to remove a name or two from your Contacts, it's easy
to do.

1. With Contacts open, tap the contact you want to delete.

2. On the information page, tap Edit.

3. On the info page that appears, scroll to the bottom of the record and
tap Delete Contact (see **Figure 7-19**).

4. When the confirmation dialog appears, tap Delete Contact to confirm
the deletion.

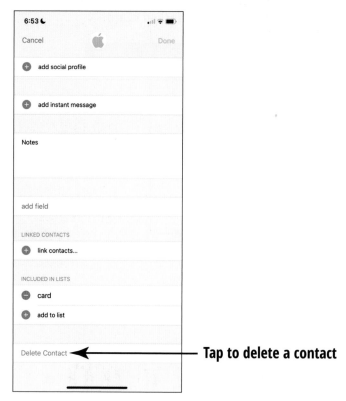

Tap to delete a contact

FIGURE 7-19

WARNING

Be careful: After you tap Delete Contact in the confirmation dialog, there's no going back! Your contact is deleted from your iPhone and any other device that syncs to your iPhone by iCloud, Google, or other means. If you change your mind during Step 4 of this process, simply tap Cancel in the upper left.

IN THIS CHAPTER

» Make, accept, and end FaceTime calls

» Switch views

» Set up an iMessage account

» Create, send, and read messages

» Send emojis and special effects

» Send audio, photos, videos, and group messages

» Hide alerts

Chapter **8**

Communicating with FaceTime and Messages

FaceTime is an excellent video-calling app that lets you call people who have FaceTime on their devices by using either a phone number or an email address. You and your friend, colleague, or family member can see each other as you talk, which makes for a much more personal and engaging calling experience.

Apple recently put a new face (user interface) on FaceTime, and adds some other tweaks and techs. Among my favorites are audio upgrades, which include spatial audio (helps those you're speaking with sound like they're right in the room, and their voice is cast in whatever direction they face in the call) and voice isolation mode (isolates your voice from other noises around you, allowing others to hear you and not your neighbor's lawnmower). I also like portrait mode, which causes your iPhone's cameras to focus on you while blurring everything behind you.

iMessage is a feature available through the preinstalled Messages app for instant messaging (IM). IM involves sending a text message to somebody's iPhone, iPod touch, Mac running macOS 10.9 or later, or iPad (using the person's phone number or email address to carry on an instant conversation). You can even send audio and video via Messages.

In this chapter, I introduce you to FaceTime and the Messages app and review their simple controls. In no time, you'll be socializing with all and sundry.

What You Need to Use FaceTime

Here's a quick rundown of the type of device and the information you need for using FaceTime's various features:

» You can use FaceTime to call people over a Wi-Fi connection who have an iPhone 4 or later, an iPad 2 or a third-generation iPad or later, a fourth-generation iPod touch or later, or a Mac (running macOS 10.6.6 or later). If you want to connect over a cellular connection, you must have an iPhone 4s or later or a third-generation cellular iPad or later.

» You can use a phone number or an email address to connect with anyone with an iPhone, an iPad, or a Mac and an iCloud account.

» The person you're contacting must have enabled FaceTime in the Settings app.

An Overview of FaceTime

FaceTime works with the iPhone's built-in cameras so that you can call other folks who have a device that supports FaceTime. You can use FaceTime to chat while sharing video images with another person. This preinstalled app is useful for seniors who want to keep up with distant family members and friends and see (as well as hear) the latest and greatest news.

You can make and receive calls with FaceTime by using a phone number or an email account and make calls to those with an iCloud account. When connected, you can show the person on the other end what's going on around you. Using the app is straightforward, although its features are limited (but getting better with each iteration).

You can use your Apple ID and iCloud account to access FaceTime, so it works pretty much right away. (See Chapter 4 for more about getting an Apple ID.)

TIP

If you're having trouble using FaceTime, make sure that the FaceTime feature is turned on. You can do so quickly: Tap Settings on the Home screen, tap FaceTime, and then tap the FaceTime switch to turn it on (green), if it isn't already. On the same screen, you can also select the phone number or email addresses or both that others can use to make FaceTime calls to you, as well as which one of those is displayed as your caller ID.

To view information for recent calls, open the FaceTime app and then tap the information icon (*i*-in-a-circle) on a recent call, and iPhone displays that person's information. You can tap the contact to call the person back.

TECHNICAL
STUFF

FaceTime is very secure, meaning that your conversations remain private. Apple encrypts (digitally protects) all of your FaceTime calls, both one-on-one and group calls, with industry-leading technology to make sure snoops are kept at bay.

Make a FaceTime Call with Wi-Fi or Cellular

If you know that the person you're calling has FaceTime available on their device, adding that person to your iPhone's contacts is a good idea so you can initiate FaceTime calls from the Contacts app if you like or from the contacts list you can access through the Face-Time app.

TIP

When you call somebody using an email address, the person must be signed in to their Apple iCloud account and have verified that the address can be used for FaceTime calls. You can access this setting by tapping Settings and then FaceTime ⇨ You Can Be Reached by FaceTime At. FaceTime for Mac users make this setting by opening the FaceTime app and selecting Preferences.

To make a FaceTime call:

1. Tap the FaceTime icon to launch the app.

   If you've made or received FaceTime calls already, you'll see a list of recent calls on the screen. You can simply tap one of those to initiate a new call, or continue with these steps to learn how to start a new call from scratch.

2. Tap the green New FaceTime button in the upper-right corner to open the New FaceTime screen. Enter a contact's name (as shown in **Figure 8-1**) by tapping the To field, or find a contact in your contacts list by tapping the green + in the upper right.

3. Tap one of the green buttons near the bottom of the screen (but above the keyboard, as shown in **Figure 8-2**) to choose a video call (green FaceTime button) or an audio call (green phone button).

   Video includes your voice and image; audio includes only your voice.

TIP

You'll see a video button if that contact's device supports FaceTime video and an audio button if the contact's device supports FaceTime audio. (If you haven't saved this person in your contacts and you know the phone number to call or email, you can just enter that information in the Enter Name, Email, or Number field.)

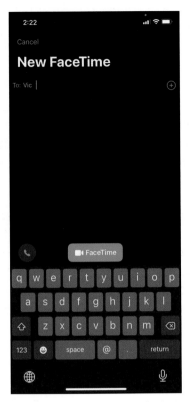

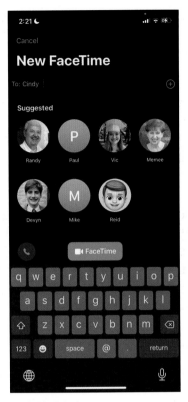

FIGURE 8-1   **FIGURE 8-2**

When the person accepts the call, you'll see a large screen that displays the recipient's image and a small screen referred to as a picture-in-picture (PiP) containing your image superimposed, as shown in **Figure 8-3**.

Want to have a little fun? Use memoji characters during your call! Memoji characters are digital illustrations that you can superimpose over your face, if your iPhone supports Face ID. During your Face-Time call, tap the effects icon, labeled in Figure 8-3 (if you don't see it, just tap the screen), tap the memoji icon in the lower-left corner (smiling illustrated character), and then select a memoji character from the list. As you can see in **Figure 8-4**, you can create your own memoji characters, as can the people you're speaking with (again, if their iPhone supports Face ID).

Effects icon

FIGURE 8-3

FIGURE 8-4

WARNING

iOS 6 and later allow you to use FaceTime over a Wi-Fi network or your iPhone cellular connection. However, if you use Face-Time over a cellular connection, you may incur costly data usage fees. To avoid the extra cost, in Settings under Cellular, toggle the FaceTime switch off (white).

Accept, Enjoy, and End a FaceTime Call

If you're on the receiving end of a FaceTime call, accepting the call is about as easy as it gets.

TIP

If you'd rather not be available for calls, you can go to Control Center and turn on a focus (probably Do Not Disturb). This stops any incoming calls or notifications other than for the people you've designated as exceptions to Do Not Disturb. After you turn on Do Not Disturb, you can use the feature's settings to schedule when it's active, allow calls from certain people, or allow a second call from the same person in a 3-minute interval to go through.

To accept, enjoy, and end a FaceTime call, follow these steps:

1. When the call comes in, tap the green button to take the call (see **Figure 8-5**).

   To reject the call, tap the red button.

TIP

2. Chat away with your friend, swapping video images.

3. To end the call, tap the red button containing the white X in the upper right (see **Figure 8-6**).

TIP

To mute sound during a call, tap the microphone icon. Tap the icon again to unmute your iPhone.

FaceTime allows group calls for up to 32 people! You can have a family reunion without leaving your front porch. To add more folks to a current call:

1. While on a call, tap the screen and then tap the *i*-in-a-circle (information icon) in the upper right to see the call options, as shown in **Figure 8-7**.

Take call   Take a still picture of the screen

Reject call   Microphone   End call

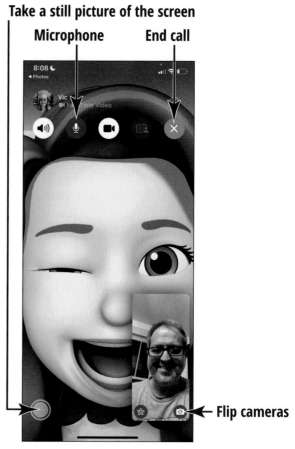

Flip cameras

FIGURE 8-5   **FIGURE 8-6**

2. Tap the Add People button.

3. Tap the To field, enter a contact or find a contact and tap the contact to add the person to the To field.

4. Tap the green Add Person to FaceTime button to add the person to the call.

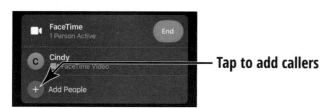

Tap to add callers

FIGURE 8-7

Switch Views

When you're on a FaceTime call, you might want to use iPhone's rear-facing camera to show the person you're talking to what's going on around you.

1. Tap the flip icon (labeled in Figure 8-6) to switch from the front-facing camera that's displaying your image to the back-facing camera that captures whatever you're looking at.

2. Tap the flip icon again to switch back to the front camera, which is displaying your image.

Set Up an iMessage Account

iMessage is a service available through the preinstalled Messages app that allows you to send an instant message (IM) to and receive an IM from others who are using an Apple iOS device, an iPadOS device, or a suitably configured Mac. iMessage is a way of sending instant messages through a Wi-Fi network, but you can send messages through your cellular connection without having iMessage activated.

TECHNICAL
STUFF

Instant messaging differs from email or tweeting in an important way. Whereas you might email somebody and wait for days or weeks before that person responds, or you might post a tweet that could sit there awhile before anybody views it, with instant messaging, communication happens almost immediately. You send an IM, and it appears on somebody's Apple device right away.

Assuming that the person wants to participate in a live conversation, the chat begins immediately, allowing a back-and-forth dialogue in real time.

1. To set up Messages, tap Settings on the Home screen.

2. Tap Messages to display the settings shown in **Figure 8-8**.

3. If iMessage isn't set on, tap the switch to toggle it on (green).

TIP

Be sure that the phone number or email account or both associated with your iPhone under the Send & Receive setting is correct. (This should be set up automatically based on your iCloud settings.) If it isn't correct, tap the Send & Receive field, tap to add an email or a phone number, and then tap Messages in the upper left to return to the previous screen.

4. To allow a notice to be sent to the sender when you've read a message, tap the switch for Send Read Receipts on. You can also choose to show a subject field in your messages, as well as many other options.

5. To leave Settings, press the Home button (or swipe up from the bottom of the screen if your iPhone doesn't have a Home button).

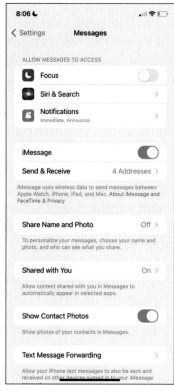
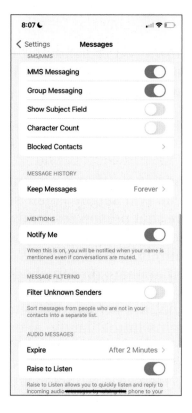

FIGURE 8-8

TIP

To enable or disable email accounts used by Messages, tap Send & Receive. Then tap an email address to enable it (check mark appears to the left) or disable it (no check mark appears to the left).

Use Messages to Address, Create, and Send Messages

Now you're ready to use Messages.

1. From the Home screen, tap the Messages button (in the dock by default).

2. Tap the compose icon (paper and pencil) in the top-right corner to begin a conversation.

3. In the form that appears (see **Figure 8-9**), begin to address your message in one of the following ways:

   - Begin to type a name in the To field, and a list of matching contacts appears.

   - Tap the dictation key (microphone icon) on the onscreen keyboard and speak the address, and then select the contact from the list that appears.

   - Tap plus (+) on the right side of the address field, and the contacts list is displayed.

4. Tap a contact in the list displayed in Step 3.

   If the contact has both an email address and a phone number, the Info dialog appears, allowing you to tap one or the other to address the message.

5. To create a message, simply tap in the message field above the keyboard (see **Figure 8-10**) and type your message.

6. To send the message, tap the send icon (labeled in Figure 8-10).

When your recipient (or recipients) responds, the conversation is displayed on the screen.

7. Tap in the message field again to respond to the last comment.

You can address a message to more than one person by simply choosing more recipients in Steps 2 and 3.

Message field   Send icon

FIGURE 8-9

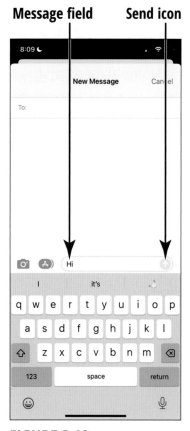

FIGURE 8-10

Read Messages

When you receive a message, it's as easy to read as email — easier, actually!

1. Tap Messages on the Home screen.

   When the app opens, you see a list of your previous text conversations.

2. Tap a conversation to see the message string, including all attachments, as shown in **Figure 8-11**.

3. To view all attachments to a message, tap the information icon (circled *i*) in the upper-right corner and scroll down. You might need to tap the person's picture to see the information icon.

FIGURE 8-11

TIP

iOS 16 introduces the cool trick of marking a conversation as read or unread. On the Messages screen, swipe a message you've already read from left to right, and then tap the blue button to mark it as unread. (A blue dot to the left indicates the message is unread.) Do the same with an unread message to mark it as read.

Clear a Conversation

When you've finished chatting, you might want to delete a conversation to remove the clutter before you start a new chat.

1. With Messages open and your conversations displayed, swipe to the left on the message you want to delete.

2. Tap the delete icon (trash can) next to the conversation you want to get rid of, as shown in **Figure 8-12**.

You can delete multiple messages at once with iOS 16! Just tap Edit in the upper-left corner of the conversation list, tap Select Messages, and then tap the circle to the immediate left of each conversation you want to delete. Once you have them all selected, tap the Delete button in the lower-right corner.

TIP

Tap the hide alerts icon (bell with a slash) to keep from being alerted to new messages in the conversation. To reactivate alerts for the conversation, swipe left again, and then tap the show alerts icon.

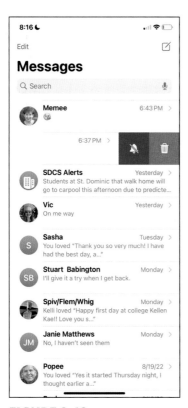

FIGURE 8-12

Send Emojis with Your Text

Emojis are small pictures that can help convey a feeling or an idea. For example, smiley faces and sad faces show emotions, and a thumbs-up emoji conveys approval.

1. From within a conversation, tap the emoji (smiley face) key on the onscreen keyboard.

   If you can't see the keyboard, tap in the Message field to display it.

2. When the emojis appear (see **Figure 8-13**), swipe left and right to find the right emoji for the moment and tap to select it.

   You can add as many as you like to the conversation.

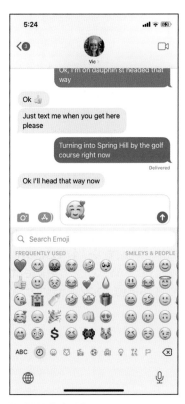

FIGURE 8-13

TIP

You can type a message and then tap the emoji key to make Messages highlight each word that has an associated emoji. You can then tap a highlighted word to insert the emoji or choose from available emojis.

Use App Drawer

App Drawer allows you to add items that spice up your messages with information from other apps installed on your iPhone, as well as drawings and other images from the web.

Tap the App Drawer icon to the left of the iMessage field in your conversation (refer to Figure 8-11). App Drawer appears (below the message field and above the keyboard). Tap an item in App Drawer to see what it offers your messaging.

App Drawer is comprised of

» **The App Store:** Tap the App Store icon, which is to the left in the App Drawer, to find tons of stickers, games, and apps for your messages.

» **Digital Touch:** Send special effects in Messages. These can include sending your heartbeat, sketching a quick picture, or sending a kiss. I explain how to use Digital Touch in a message shortly.

» **Other apps you have installed:** These may appear if they have the capability to add functions and information to your messages. For example, use ESPN to send the latest scores, use YouTube to send a link to a video (as I'm doing in **Figure 8-14**), use an AccuWeather app icon to let your friend know what the weather's like nearby, or use the Fandango app to send movie information.

FIGURE 8-14

Digital Touch is one of the most personal ways to send special effects to others. Here's a close look:

1. To send a Digital Touch in a message, open a conversation and tap the Digital Touch icon, shown in **Figure 8-15**.

2. In the Digital Touch window, tap the gray expansion handle (labeled in Figure 8-15) to open the full window.

3. Tap the information icon in the lower right (*i* in a gray circle), and you'll see a list of the gestures and what they do.

4. Perform a gesture in the Digital Touch window and it will go to your recipient, as shown in **Figure 8-16**.

Expansion handle   **Digital Touch button**

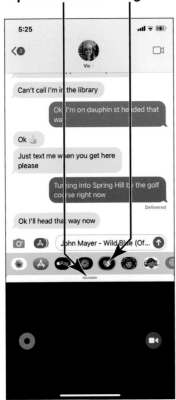

FIGURE 8-15

FIGURE 8-16

Send and Receive Audio

When you're creating a message, you can also create an audio message.

1. With Messages open, tap the new message icon (paper and pencil) in the top-right corner.

2. Enter an addressee's name in the To field.

3. Tap the audio icon (white waveform in a blue oval) in App Drawer and a red microphone appears at the bottom (see **Figure 8-17**).

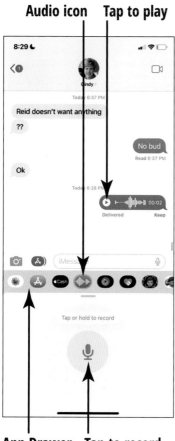

Audio icon   Tap to play

App Drawer   Tap to record

FIGURE 8-17

4. Tap the microphone while you speak your message or record a sound or music near you.

5. When you've finished recording, tap the microphone again.

6. To send your message, tap the send icon (upward-pointing arrow in the speech bubble, just above the play icon).

   The message appears as an audio track in the recipient's Messages inbox (refer to Figure 8-17). To play the track, the recipient just holds the phone up to an ear or taps the play icon.

Send a Photo or Video

When you're creating a message, you can also send a picture or create a short video message.

1. With Messages open, tap the new message icon (paper and pencil) in the top-right corner.

2. Tap the camera icon to the left of the message field to open the Camera app, and then take a picture.

3. After you take the picture, the tools shown in **Figure 8-18** appear and you can work with the picture before sending it along to your recipient:

   - Tap Retake to take a different picture.

   - Tap Effects to add items like memoji or customized text, or to apply filters.

   - Tap Edit to edit the picture.

   - Tap Markup to add notes or other text to your picture.

   - Tap Done to place the picture in your message but not send it yet.

   - Tap the send icon (upward-pointing arrow in the blue circle) to send the picture immediately.

FIGURE 8-18

Understand Group Messaging

If you want to start a conversation with a group of people, you can use group messaging. Group messaging is great for keeping several people in a conversational loop.

Group messaging functionality includes the following features:

» When you participate in a group message, you see all participants in the information screen for the message. Access the information screen by tapping the icons for your contacts at the top of the message thread. You can drop people whom you don't want to include any longer and leave the conversation yourself when you want to by simply tapping Info and then tapping Leave this Conversation.

» When you turn on Hide Alerts in the information screen of a message, you won't get notifications of messages from this group, but you can still read the group's messages at a later time (this also works for individual messages).

Taking you further into the workings of group messages is beyond the scope of this book, but if you're intrigued, go to `https://support.apple.com/en-us/HT202724` for more information.

Activate the Hide Alerts Feature

If you don't want to get notifications of new messages from an individual or a group for a while, you can use the Hide Alerts feature.

1. With a message open, tap open the information screen by tapping the icon of the contact(s) at the top of the screen.

2. Tap the Hide Alerts switch to turn on the feature.

3. To turn the feature off later, return to Details and tap the Hide Alerts switch again.

IN THIS CHAPTER

» **Use the Calculator app**

» **Get your bearings with Compass**

» **Record voice memos**

» **Measure and level**

» **Get started with the Home app**

» **Translate words and phrases**

Chapter **9**

Using Handy Utilities

U tilities are simple apps that can be very useful indeed to help with common tasks, such as calculating a meal tip, finding your way on a hike in the woods, or translating phrases from one language to another.

In this chapter, you find out how to use the Calculator app to keep your numbers in line. You also explore a few other apps: Compass to help you find your way, Voice Memos so that you can record your best ideas for posterity (perhaps while you're on a hike using the Compass app for direction), Measure to help determine distances (you can find out just how far of a jump it is over the creek in your path), Home to begin automating your home's tech (the coffee will be hot and ready when you get back), and Translate to assist with conversing in languages you don't natively speak or are still learning (you never know the interesting folks you might mean on the trail).

Use the Calculator App

This one won't be rocket science. The Calculator app works like just about every calculator app (or actual calculator, for that matter) you've ever used. Follow these steps:

1. Tap the Calculator app icon (shown in **Figure 9-1**) to open it.

   You'll find the app in the Utilities folder.

2. Tap a few numbers (see **Figure 9-2**), and then use any of these functions and additional numbers to perform calculations:

   - **+, –, ×, and ÷:** These familiar buttons add, subtract, multiply, and divide, respectively, the number you've entered.

   - **+/–:** If the calculator is displaying a negative result, tap this to change it to a positive result, and vice versa.

FIGURE 9-1

FIGURE 9-2

- **AC/C:** This is the clear button; its name will change depending on whether you've entered anything. (AC clears all; C clears just the last entry after you've made several entries.)

- **=:** This button produces the result of whatever calculation you've entered.

If you have a true mathematical bent, you'll be delighted to see that if you turn your phone to landscape orientation, you'll get additional features that turn the basic calculator into a scientific calculator (see **Figure 9-3**). Now you can play with calculations involving cosines, square roots, tangents, and other fun stuff. You can also use memory functions to work with stored calculations.

FIGURE 9-3

Find Your Way with Compass

Compass is a handy tool for figuring out where you are, assuming that you get cellular reception wherever you are. To find directions with Compass, follow these steps:

1. In the Utilities folder, tap the Compass icon to get started.

   The first time you do this, if you haven't enabled location access, a message appears asking whether your iPhone can use your current location to provide information. When prompted, tap either Allow Once, Allow While Using App (recommended), or Don't Allow.

2. If you are using Compass for the first time and are asked to tilt the screen to roll a little red ball around a circle, do so.

This exercise helps your iPhone calibrate the Compass app.

3. When the Compass app appears (see **Figure 9-4**), move around with your iPhone, and the compass will indicate your current orientation in the world.

4. To lock in your heading, tap the bold white line indicating your current direction.

As you move, a red arc appears, as shown in **Figure 9-5**, to indicate how far off course you've deviated.

FIGURE 9-4

FIGURE 9-5

TECHNICAL
STUFF

The Compass app can use either true north or magnetic north for navigation. *True north* refers to the direction you follow from where you are to get to the North Pole. *Magnetic north* is correlated relative to the Earth's magnetic field. True north is the more accurate measurement because of the tilt of the Earth. By default, Compass uses magnetic north, but you can change this by going to Settings ⇨ Compass and toggling the Use True North switch on (green).

Record Voice Memos

The Voice Memos app allows you to record memos, trim them, share them by email or instant message with Messages, synchronize recordings and edits across Apple devices (iPhone, iPad, and Mac), and label recordings so that you find them easily. iOS 16 also enables you to organize recordings by creating folders in the app, and to speed playback so that it doesn't take as long to listen to your recordings.

TIP

If you use iCloud with Voice Memos, memos you record on your iPhone, iPad, or Mac will sync with all your Apple devices (if they're signed into iCloud with your Apple ID). Bear this in mind when you want to keep your memos private but share your devices.

To record voice memos, follow these steps:

1. Tap the Voice Memos icon (in the Utilities folder) to open the app.

2. In the Voice Memos app, tap All Recordings or a folder, and then tap the red record icon at the bottom of the screen (shown in **Figure 9-6)** to record a memo.

   The record icon changes to a red stop icon (square in a circle) when you're recording. A red waveform moving from right to left indicates that you're in recording mode, as shown in **Figure 9-7**.

3. Swipe up on the recording window to reveal the advanced controls, as shown in **Figure 9-8**.

   Now the waveform is moving left to right, indicating the recording is continuing.

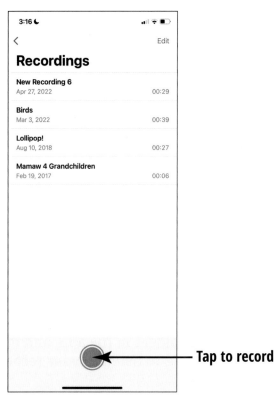

Tap to record

FIGURE 9-6   FIGURE 9-7

4. While recording:

   (a) Tap the name of the recording (called New Recording by default) to give it a more descriptive name.

   (b) Tap the red pause icon to pause the recording, and then tap Resume to continue recording. While paused, you can also tap the play icon to play what you've recorded so far, and then tap Resume to continue recording.

   (c) While paused, drag the waveform to a place in the recording you'd like to record over, and then tap the Replace button to begin recording from there.

   (d) Tap Done in the lower-right corner to stop recording. The new recording will appear in the All Recordings list or the folder you were in when you began recording or both.

5. Tap a recording in the All Recordings list to open its controls. From here you can

- Tap the play icon to play back the recording.
- Tap the forward or reverse icons to move forward or backward, respectively, 15 seconds in the recording.
- Tap the trashcan icon to delete the memo.
- Tap the name of the recording to rename it.

6. Tap the options icon (labeled in **Figure 9-9**) in the recording playback controls. Your options are

- *Playback Speed:* Drag the slider to slow down or speed up playback.

Options icon

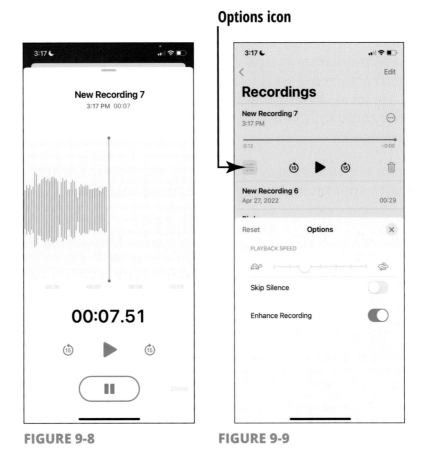

FIGURE 9-8   FIGURE 9-9

- *Skip Silence:* Toggle this switch on (green) to have gaps in the audio removed.

- *Enhance Recording:* Toggle this switch on to remove background noise from the recording.

TIP

Deleted voice memos are kept for 30 days in the Recently Deleted folder in the Voice Memos list. You can retrieve a deleted memo by tapping the Recently Deleted button, tapping the name of the memo you want to retrieve, and then tapping Recover.

Measure Distances and Level Objects

iOS 16 uses the latest advancements in AR (augmented reality) and your iPhone to offer you a cool way to ditch your measuring tape and level: the Measure app! This app enables you to use your iPhone to measure distances and objects simply by pointing your iPhone at them. It also helps you check that things (such as pictures and whatnot) are level. Measure is fun to play with and surprisingly accurate to boot (although you still may want to hang onto your trusty measuring tape).

TIP

To increase the accuracy of your measurements when using the Measure app, make sure you have plenty of light.

1. Open the Measure app by tapping its icon (in the Utilities folder) and then tap the Measure button in the lower left to use the measurement feature.

2. When prompted (see **Figure 9-10**), calibrate the Measure app by panning your iPhone around so that the camera gets a good look at your surroundings.

3. Add the first reference point for your measurement by aiming the white targeting dot in the center of the screen to the location of your first reference point, as shown in **Figure 9-11**. Tap the add a point icon (white circle with gray +) to mark the point.

4. Mark the second reference point by placing the targeting dot on the location (see **Figure 9-12**) and tapping the add a point icon again.

FIGURE 9-10   **FIGURE 9-11**

TIP

Should you make a mistake or simply want to start afresh, tap the Clear button in the upper-right corner to clear your reference points and begin again.

The length of your measurement is displayed as a white line, with the distance is shown in the middle.

5. Continue to make measurements by aiming the targeting dot at a previous reference point, tapping the add a point icon, and moving your iPhone to the next reference point, where you again tap the add a point icon to make a new measurement, as shown in **Figure 9-13**.

6. To save an image of your measurements to the Recents album in the Photos app, tap the white capture icon (to the right of the add a point icon in portrait mode, and above the add a point icon in landscape mode).

FIGURE 9-12   **FIGURE 9-13**

Here's how to determine whether an object or a surface is level. Let's say you're trying to figure out if a picture frame is perfectly straight on your wall:

1. On the main screen of the Measure app, tap the level icon in the lower right.

   The screen displays how many degrees from zero that the surface is, as shown in **Figure 9-14**.

2. Align your iPhone along the upper border of the frame.

   If the surface beneath your phone is level, the screen turns green, and you'll see 0 degrees in the middle of the screen. If the frame is a little crooked, you might see 1 or 2 degrees on the iPhone screen.

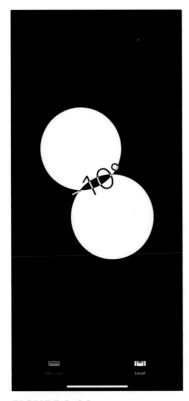

FIGURE 9-14

TIP

The level feature of the Measure app can come in handy when working on construction projects. While a conventional level is typically best, I've successfully used the Measure app several times when one wasn't close by.

Discover the Home App

Since the smart-home movement began a few years ago, controlling your home remotely meant juggling several apps: one for your lights, one for your garage doors, one for your thermostat, one for your oven, and on and on. While some developers have tried to create apps that worked with multiple smart-home platforms by multiple manufacturers, none had the clout or the engineering workforce to pull things together — until Apple jumped in.

The Home app on your iPhone works with multiple smart-home platforms and devices, enabling you to control compatible smart devices in your home from one easy-to-use app.

Here's a list of the types of devices you can control remotely using your iPhone, as long as you have at least a cellular data connection: lighting, locks, windows and window shades, heating and cooling systems, speakers, humidifiers and air purifiers, security systems, garage doors, plugs and switches, sensors, video cameras, smoke and carbon monoxide detectors, and more!

TIP

If you want to use the Home app with your smart-home devices, make sure you see the Works with Apple HomeKit symbol on the packaging or on the website (if you purchase the device online). Apple has an ever-growing list of HomeKit-enabled devices at www.apple.com/ios/home/accessories. You can also buy HomeKit-enabled devices on Apple's website: www.apple.com/shop/accessories/all/homekit.

Because there are so many ways to configure and use the Home app and so many different accessories you can control with it, it's beyond the scope of this book to cover the app in detail. Apple offers a great overview at www.apple.com/ios/home.

Translate Words and Phrases

I remember thinking how cool the Universal Translator was the first time I saw it in the original *Star Trek* television series. The *Enterprise* crew would meet a being that spoke a language unknown to them, but the Universal Translator would have them swapping jokes in no time flat. With the introduction of the Translate app, it would seem that Apple envisions the iPhone as a first step towards such a device in the early 21st century.

Translate allows you to, well, translate words and phrases from one language into another, supporting 18 languages. Translate will even help you engage in conversations on the fly with its conversation mode.

TIP

Translate currently supports English, Spanish, Mandarin Chinese, Arabic, Brazilian Portuguese, Russian, Korean, Italian, German, French, Turkish, Dutch, Thai, Polish, Vietnamese, Indonesian, and Japanese.

To start translating:

1. Tap the Translate app icon to open it.

2. Tap the down arrow next to the button in the upper left (shown in **Figure 9-15**), and then tap to select the language you want translated.

3. Tap the down arrow next to the button in the upper right and tap to select the language you want your words or phrases translated to.

4. Enter your words or phrases either by tapping the Enter Text area (near the bottom of the screen) and typing or pasting your text, or by tapping the blue microphone icon and speaking your text.

   Translate displays your original text in white or black (depending on whether your phone is using light or dark mode) and the translated text in blue, as shown in **Figure 9-16**.

5. To hear the word or phrase spoken, tap the blue play icon under the translated text.

6. To save the translation to your favorites, which is helpful if the translation is a common phrase you'll need to refer to often, tap the small star under the translated text. To access your Favorites list, just tap the Favorites icon at the bottom right of the Translate app's window.

7. If you'd like to see a comprehensive definition of a word, complete with usage examples, tap a word in the translation to highlight it, and then tap the dictionary icon below it.

TIP

Tap the Camera icon at the bottom of the app to allow Translate to translate text from images. This feature is handy when trying to translate road signs, restaurant menus, books, you name it!

Select a language to translate from   **Select a language to translate to**

Microphone

FIGURE 9-15

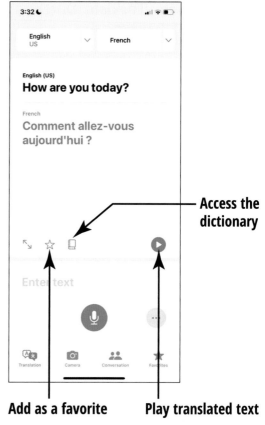

Access the dictionary

Add as a favorite   **Play translated text**

FIGURE 9-16

Conversation mode allows you to carry on a conversation with someone who speaks a different language, and you both can see the translations in real time. Here's how to make that happen:

1. With the Translate app open, simply tap the Conversation icon at the bottom of the screen and conversation mode will open, as shown in **Figure 9-17**.

   By default, Translate will automatically detect the language being spoken by each participant in the conversation.

FIGURE 9-17

2. Take turns tapping the microphone icon and speaking.

When the speaker stops, the word or phrase is translated so that the other participant can read what the speaker said.

3. Tap the orientation icon (labeled in **Figure 9-18**) to decide whether you and the person you're speaking with would like to translate the conversation side by side (the default shown in Figure 9-18) or face to face (shown in **Figure 9-19**).

TIP

If the language isn't automatically detected correctly or if the translation isn't quite right, disabling a few settings might help. Tap the more icon (labeled in Figure 9-18) and then tap Auto Translate or Detect Language or both to enable or disable the features.

FIGURE 9-18

FIGURE 9-19

IN THIS CHAPTER

» Set brightness and wallpapers

» Set up and use VoiceOver

» Use iPhone with hearing aids

» Set up subtitles, captioning, and other hearing settings

» Learn with Guided Access

» Control your iPhone with your voice

Chapter **10**

Making Your iPhone More Accessible

Phone users are a diverse group, and some face visual, motor, or hearing challenges. If you're one of these folks, you'll be glad to know that Apple offers some handy accessibility features for your iPhone. You can make your screen easier to read as well as set up the VoiceOver feature to read onscreen elements out loud. Voice Control, Numbers, and Grids are welcome accessibility features to help you navigate more easily. Then there are a slew of features you can turn on or off, including Zoom, Invert Colors, Speak Selection, and Large Type.

If hearing is your challenge, you can do the obvious and adjust the system volume. If you wear hearing aids, you can choose the correct settings for using Bluetooth or another hearing aid mode.

The iPhone also has features that help you deal with physical and motor challenges. And the Guided Access feature helps if you have difficulty focusing on one task. All of these accessibility features and more are covered in this chapter.

Set Brightness

Especially when using iPhone as an e-reader, you may find that a slightly less-bright screen reduces strain on your eyes. To manually adjust screen brightness, follow these steps:

1. Tap the Settings icon on the Home screen.

TIP

If glare from the screen is a problem for you, consider getting a screen protector. This thin film both protects your screen from damage and reduces glare. You can easily find them on Amazon and at just about any cellphone dealer.

2. In Settings, go to Accessibility ⇨ Display & Text Size.

3. Tap the Auto-Brightness on/off switch at the bottom of the screen (see **Figure 10-1**) to turn off this feature (the switch turns white when off).

4. Tap Back (or possibly Accessibility, depending on your iPhone model and settings) in the upper-left corner and then tap Settings in the same location.

5. Tap Display & Brightness and then tap and drag the Brightness slider (refer to **Figure 10-2**) to the right to make the screen brighter or to the left to make it dimmer.

6. Press the Home button (or swipe up from the bottom of the screen for iPhone models without a Home button) to close Settings.

TIP

In the Apple Books e-reader app, you can set a custom themes for your book pages, which might be easier on your eyes. See Chapter 17 for more about using Apple Books.

FIGURE 10-1

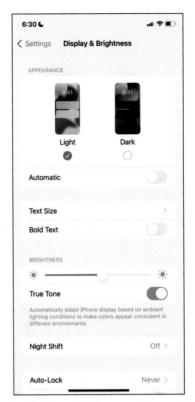

FIGURE 10-2

Change the Wallpaper

You might like the default iPhone background image on your iPhone, but it may not be the one that works best for you. Choosing a different wallpaper may help you more easily see all the icons on your Home screen. Follow these steps:

1. Tap the Settings icon on the Home screen.

2. In Settings, tap Wallpaper.

3. In the Wallpaper settings, tap Add New Wallpaper.

4. Tap a wallpaper category (see **Figure 10-3**) to view your choices. Tap a sample to select it.

TIP

If you prefer to use a picture that's on your iPhone, tap an album in the lower part of the Wallpaper screen to locate a picture; and then tap the picture to use it as your wallpaper.

5. In the preview that appears (see **Figure 10-4**), tap any options on the screen that you'd like to customize (what you see depends on the wallpaper you're previewing), and then tap Add in the upper-right corner to select the new wallpaper.

6. Back on the Wallpaper screen, where your new wallpaper has been applied to your lock and Home screens, tap the Customize button under either if you'd like to make any changes.

7. Press the Home button, or swipe up from the bottom of the screen if your iPhone doesn't have a Home button.

FIGURE 10-3

FIGURE 10-4

You return to your Home screen with the new wallpaper set as the background. And the next time you view your lock screen, the new wallpaper should appear there, too.

Set Up VoiceOver

VoiceOver reads the names of screen elements and settings to you, but it also changes the way you provide input to the iPhone. In Notes, for example, you can have VoiceOver read the name of the Notes buttons to you, and when you enter notes, it reads words or characters that you've entered. It can also tell you whether such features as Auto-Correction are on.

VoiceOver is even smarter in iOS 16 than in previous incarnations. It includes support for apps and websites that may not have built-in accessibility support. It can read descriptions of images in apps and on the web, and it can identify and speak text it finds in images.

To turn on VoiceOver, follow these steps:

1. Tap the Settings icon on the Home screen.

2. In Settings, tap Accessibility.

3. In the Accessibility pane, shown in **Figure 10-5**, tap VoiceOver.

4. In the VoiceOver pane, shown in **Figure 10-6**, tap the VoiceOver on/off switch to turn on this feature (the button turns green).

   With VoiceOver on, you must first single-tap to select an item such as a button, which causes VoiceOver to read the name of the button to you. Then you double-tap the button to activate its function.

5. Tap the VoiceOver Practice button to select it, double-tap the button to open VoiceOver Practice, and then tap VoiceOver Practice in the middle of the screen to begin. Practice using gestures (such as pinching and flicking left), and VoiceOver tells you what action each gesture initiates.

6. To return to the VoiceOver dialog, tap the Done button and then double-tap the same button.

Tap here

Tap here

FIGURE 10-5

FIGURE 10-6

7. Tap the Verbosity button once and then double-tap to open its options:

- Tap the Speak Hints on/off switch and then double-tap the switch to turn the feature on (or off). VoiceOver speaks the name of each tapped item.

- Tap once and then double-tap the VoiceOver button in the upper-left corner to go back to the VoiceOver screen.

8. If you would like VoiceOver to speak descriptions of images in apps or on the web, scroll down and then tap and double-tap VoiceOver Recognition, tap and double-tap Image Descriptions, and tap and double-tap the Image Descriptions switch to toggle the setting on.

TIP

Don't ignore the Sensitive Content Output section of the Image Descriptions page. If you don't want everyone in the room to hear the content of an image, select the Play Sound setting or the Do Nothing setting rather than the Speak setting.

9. Tap and double-tap the Back button in the upper-left corner, and then do the same thing again to return to the main VoiceOver screen.

10. If you want VoiceOver to read words or characters to you (for example, in the Notes app), scroll down using a three-finger swipe and then tap and double-tap Typing, and then tap and double-tap Typing Feedback.

11. In the Typing Feedback dialog, tap and then double-tap to select the option you prefer in both the Software Keyboards section and the Hardware Keyboards section.

The Words option causes VoiceOver to read words to you but not characters, such as the dollar sign ($). The Characters and Words option causes VoiceOver to read both, and so on.

12. Press the Home button or swipe up from the bottom of the screen (for iPhones with no Home button) to return to the Home screen.

The following section shows how to navigate your iPhone after you've turned on VoiceOver.

TIP

You can use the Accessibility Shortcut setting to help you more quickly turn the VoiceOver, Zoom, Switch Control, AssistiveTouch, Grayscale, and Invert Colors features on and off:

1. In the Accessibility screen, tap Accessibility Shortcut.

2. Choose what you want three presses of the Home button (or side button for iPhones without a Home button) to activate.

Now three presses with a single finger on the Home button or side button provides you with the option you selected.

Use VoiceOver

After VoiceOver is turned on, you need to figure out how to use it. Using it is awkward at first, but you'll get the hang of it!

Here are the main onscreen gestures you should know how to use:

» **Tap an item to select it.** VoiceOver speaks its name.

» **Double-tap the selected item.** The item is activated.

» **Flick three fingers.** The page scrolls.

TIP If your iPhone has a Home button, simply press it to unlock — simple. If your iPhone doesn't have a Home button, look at your iPhone (for Face ID to recognize you) and then slowly move your finger up from the bottom of the screen until you hear two tones or feel a vibration.

TIP If tapping with two or three fingers is difficult, try tapping with one finger from one hand and one or two from the other. When double- or triple-tapping, you have to perform these gestures as quickly as you can for them to work.

Table 10-1 provides additional gestures to help you use VoiceOver. If you want to use this feature often, I recommend the VoiceOver section of the online *iPhone User Guide,* which goes into great detail about using VoiceOver. You'll find the *User Guide* at `https:// support.apple.com/manuals/iphone`. When there, just click the model of iPhone or the version of iOS you have to read its manual. You can also get an Apple Books version of the manual through that app in the Book Store. (See Chapter 17 for more information.)

TABLE 10-1   VoiceOver Gestures

Gesture	Effect
Flick right or left.	Select the next or preceding item.
Tap with two fingers.	Stop speaking the current item.
Flick two fingers up.	Read everything from the top of the screen.
Flick two fingers down.	Read everything from the current position.
Flick three fingers up or down.	Scroll one page at a time.
Flick three fingers right or left.	Go to the next or preceding page.
Tap three fingers.	Speak the scroll status (for example, line 20 of 100).
Flick four fingers up or down.	Go to the first or last element on a page.
Flick four fingers right or left.	Go to the next or preceding section (as on a web page).

Customize Vision Settings

Several Vision features are simple on/off settings that you can turn on or off after you tap Settings ➪ Accessibility:

» **Zoom:** The Zoom feature enlarges the contents displayed on the iPhone screen when you double-tap the screen with three fingers. The Zoom feature works almost everywhere in iPhone: in Photos, on web pages, on your Home screens, in your Mail, in Music, and in Videos — give it a try!

» **Display & Text Size:** Includes such features as Color Filters (aids those with color blindness), Reduce White Point (reduces the intensity of bright colors), and Invert colors (reverse colors on your screen so that white backgrounds are black and black text is white) using either Classic Invert or Smart Invert (reverses colors for everything except images and other media files).

» **Spoken Content:** Options here include the ability to have your iPhone speak items you've selected or hear the content of an entire screen, highlight content as it's spoken, and more.

- » **Larger Text (under Accessibility ⇨ Display & Text Size):** If having larger text in such apps as Contacts, Mail, and Notes would be helpful, you can turn on the Larger Text feature and choose the text size that works best for you.

- » **Bold Text (under Accessibility ⇨ Display & Text Size):** Turning on this setting restarts your iPhone (after asking for permission to do so) and then causes text in various apps and in Settings to be bold.

- » **Button Shapes (under Accessibility ⇨ Display & Text Size):** This setting applies shapes to buttons so they're more easily distinguishable.

- » **Reduce Transparency (under Accessibility ⇨ Display & Text Size):** This setting helps increase legibility of text by reducing the blurs and transparency effects that make up a good deal of the iPhone user interface.

- » **Increase Contrast (under Accessibility ⇨ Display & Text Size):** Use this setting to set up backgrounds in some areas of iPhone and apps with greater contrast, which should improve visibility.

- » **On/Off Labels (under Accessibility ⇨ Display & Text Size):** If you have trouble making out colors and therefore find it hard to tell when an on/off switch is on (green) or off (white), use this setting to add a circle to the right of a switch when it's off and a white vertical line to a switch when it's on.

- » **Reduce Motion (under Accessibility ⇨ Motion):** Tap this accessibility feature and then tap the on/off switch to turn off the parallax effect, which causes the background of your Home screens to appear to float as you move the phone around.

Use iPhone with Hearing Aids

If you have Bluetooth enabled or use another style of hearing aid, your iPhone may be able to detect it and work with its settings to improve sound on your phone calls. Follow these steps to connect your hearing aid to your iPhone.

TIP Your hearing aids may come with their own instructions, and probably even an app, for setting up and configuring options. If so, use those instead of what I discuss here.

1. Tap Settings on the Home screen.

2. Tap Accessibility and then scroll down to the Hearing section and tap Hearing Devices.

   On the following screen, your iPhone searches for hearing-aid devices.

TIP If you have a non-MFi (Made for iPhone) hearing aid, add your hearing aid in Bluetooth settings. To do so, go to Settings ⇨ Bluetooth, make sure the Bluetooth toggle switch is on (green), and select your hearing aid in the list of devices.

3. When your device appears, tap it.

4. Tap the Back button in the upper-left corner of the screen, scroll back down to the Hearing Devices section (if you're not automatically returned there), and tap the Hearing Aid Compatibility switch on to possibly improve audio quality when you're using your hearing aid.

Adjust the Volume

Though individual apps (such as Music and Video) have their own volume settings, you can set your iPhone system volume for your ringer and alerts as well to help you better hear what's going on. Follow these steps:

1. Tap Settings on the Home screen and then tap Sounds & Haptics.

2. In the Sounds & Haptics settings that appear (see **Figure 10-7**), tap and drag the Ringtone and Alert Volume slider to adjust the volume of these audible attention grabbers:

   - Drag to the right to increase the volume.
   - Drag to the left to lower the volume.

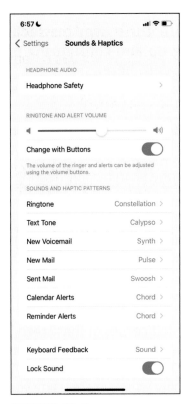

FIGURE 10-7

TIP

In the Sounds & Haptics settings, you can turn on or off the sounds that iPhone makes when certain events occur (such as receiving new Mail or Calendar alerts). These sounds are turned on by default.

3. To return to the Home screen, press the Home button or swipe up from the bottom of the screen (iPhone models without a Home button).

TIP

Even those with perfect hearing sometimes have trouble hearing a phone ring, especially in busy public places. To have your phone vibrate when a call is coming in, modify the Vibration settings under each item in the Sounds and Haptic Patterns section.

Set Up Subtitles and Captioning

Closed captioning and subtitles help folks with hearing challenges enjoy entertainment and educational content. Follow these steps:

1. Tap Settings on the Home screen, and then tap Accessibility.

2. Scroll down to the Hearing section and tap Subtitles & Captioning.

3. On the following screen, tap the Closed Captions + SDH (Subtitles for the Deaf and Hard of Hearing) switch on (green).

   You can also tap Style and choose a text style for the captions, as shown in **Figure 10-8**. A neat video helps show you what your style will look like when the feature is in use. Tap the black box in the lower right of the video to expand it to full screen. Tap the screen to return to the Style screen.

4. To return to the Home screen, press the Home button or swipe up from the bottom of the screen (for iPhone models without a Home button).

FIGURE 10-8

Manage Other Hearing Settings

Several hearing accessibility settings are simple on/off settings, including

» **Mono Audio (under Accessibility ⇨ Audio/Visual):** Using the stereo effect in headphones or a headset breaks up sounds so that you hear a portion in one ear and a portion in the other ear. The purpose is to simulate the way your ears process sounds. However, if you're hard of hearing or deaf in one ear, you're hearing only a portion of the sound in your hearing ear, which can be frustrating. If you have such hearing challenges and want to use iPhone with a headset connected, you should turn on Mono Audio. When it's turned on, all sound is combined and distributed to both ears. You can use the Balance slider below Mono Audio to direct sounds to provide more hearing balance.

» **LED Flash for Alerts (the last option under Accessibility ⇨ Audio/Visual):** If you need a visual cue when an alert is spoken, turn this setting on.

» **Phone Noise Cancellation (under Accessibility ⇨ Audio/Visual):** If you're annoyed at ambient noise when you make a call in public (or noisy private) settings, turn on the Phone Noise Cancellation feature (iPhone 5 or later). When you hold the phone to your ear during a call, this feature reduces background noise to some extent.

» **RTT/TTY:** RTT stands for Real-Time Text, which provides better text support during calls. TTY is a symbol indicating teletype machine capabilities. The iPhone is compatible with teletype machines via the iPhone TTY adapter, which can be purchased from Apple. The TTY option allows you to enable either Software TTY, Hardware TTY, or both. RTT and TTY are available for your iPhone only if your cell carrier supports one or both of them.

Turn On and Work with AssistiveTouch

If you have difficulty using buttons, the AssistiveTouch menu aids input using the touchscreen.

1. To turn on AssistiveTouch, tap Settings on the Home screen and then tap Accessibility.

2. In the Accessibility pane, tap Touch and then tap AssistiveTouch. In the pane that appears, tap the on/off switch for AssistiveTouch to turn it on (see **Figure 10-9**).

   A dark circle, called the AssistiveTouch menu button, then appears on the right side of the screen (although you're unable to see it in Figure 10-9; it's there when the feature is enabled). This menu button now appears in the same location in whatever apps you display on your iPhone, though you can move it around with your finger.

3. Tap the AssistiveTouch menu to display the options shown in **Figure 10-10**.

   The panel offers Notification Center and Control Center options.

4. You can tap Custom or Device on the panel to see additional choices, tap Siri to activate the personal assistant feature, tap Notification Center or Control Center to display those panels, or tap Home to go directly to the Home screen.

5. To return to the Home screen after you've chosen an option, press the Home button or swipe up from the bottom of the screen (for iPhones without a Home button).

Table 10-2 shows the major options available in the AssistiveTouch menu and their purpose.

FIGURE 10-9

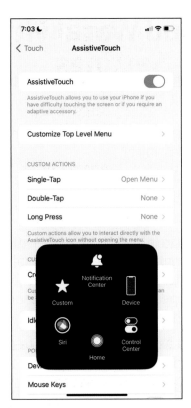

FIGURE 10-10

TABLE 10-2   AssistiveTouch Options

Option	Purpose
Siri	Activates the Siri feature, which allows you to speak questions and make requests of your iPhone.
Custom	Displays a set of gestures, with pinch and rotate, long-press, double tap, and hold and drag gestures preset; you can tap any of the other blank squares to add your own favorite gestures.
Device	Displays presets that enable you to rotate the screen, lock the screen, turn the volume up or down, mute or unmute sound, and more.
Home	Sends you to the Home screen.
Control Center	Opens Control Center.
Notification Center	Opens Notification Center with reminders, Calendar appointments, and so on.

Turn On Additional Physical and Motor Settings

Use these on/off settings in the Accessibility settings to help you deal with how fast you tap and how you answer phone calls:

» **Home Button (appears only if your iPhone does indeed have a Home button):** Sometimes if you have dexterity challenges, it's hard to double-press or triple-press the Home button fast enough. Choose the Slow or Slowest option for this setting to allow you a bit more time to make that second or third press. Also, the Rest Finger to Open feature at the bottom of the screen is helpful because it allows you to simply rest your finger on the Home button — as opposed to pressing it — to open your iPhone using Touch ID (if enabled).

» **Call Audio Routing (under Accessibility ⇨ Touch):** If you prefer to use your speaker phone to receive incoming calls, or you typically use a headset that allows you to tap a button to receive a call, tap this option and then choose Bluetooth Headset or Speaker. Speakers and headsets can both provide a better hearing experience for many.

TIP If you have certain adaptive accessories that allow you to control devices with head gestures, you can use them to control your iPhone, highlighting features in sequence and then selecting one. Use the Switch Control feature in the Accessibility settings to turn this mode on and make your selections.

Focus Learning with Guided Access

Guided Access is a feature that you can use to limit a user's access to iPhone to a single app, and even limit access in that app to certain features. This feature is useful in several settings, ranging from a classroom, for use by someone with attention deficit disorder, and even to a public setting (such as a kiosk where you don't want users to be able to open other apps).

1. Tap Settings and then tap Accessibility.

2. Scroll down and tap Guided Access. On the screen that appears, toggle the Guided Access switch to turn the feature on (green).

3. Tap Passcode Settings and then tap Set Guided Access Passcode.

   You can set a passcode so that those using an app can't return to the Home screen to access other apps. You can also activate Touch ID (only iPhones with a Home button) or Face ID (only iPhone models without a Home button) to perform the same function, as shown in **Figure 10-11**.

4. In the Set Passcode dialog that appears, enter a passcode using the numeric pad. Enter the number again when prompted.

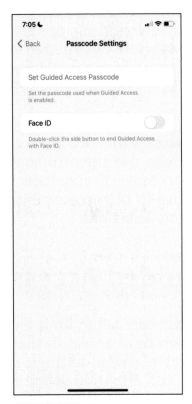

FIGURE 10-11

5. Press the Home button or swipe up from the bottom of the screen (iPhones without a Home button), and tap an app to open it.

6. Rapidly press the Home button (side button for iPhones without a Home button) three times. If the Guided Access screen doesn't open automatically, tap the Guided Access button at the bottom of the screen.

7. Tap Options at the bottom to display the following options and make your selections:

   - *Sleep/Wake Button or Side Button:* You can put your iPhone to sleep or wake it up with three presses of the Home button or side button, depending on your iPhone model.

   - *Volume Buttons:* Tap to toggle this switch on or off. You might use this setting, for example, if you don't want users to be able to adjust the volume using the volume toggle on the side of the iPhone.

   - *Motion:* Turn this setting off if you don't want users to move the iPhone around — for example, to play a race car driving game.

   - *Keyboards:* Use this setting to prohibit people using this app from entering text with the keyboard.

   - *Touch:* If you don't want users to be able to use the touchscreen, turn this off.

   - *Time Limit:* Tap to toggle this option on or off and use the displayed settings to set a time limit for the use of the app.

8. Tap Done to hide the options.

   At this point, you can also use your finger to circle areas of the screen that you want to disable, such as a Store button in the Music app.

9. Tap the Start button (upper-right corner).

   Guided Access is now live, and users can explore the app within the limits you set.

10. When the user has finished with Guided Access, press the Home button (side button for iPhones with no Home button) three times. Enter your passcode, if you set one, and tap End.

11. To return to the Home screen, press the Home button or swipe up from the bottom of the screen (iPhones without a Home button).

One-Handed Keyboard

The one-handed keyboard makes typing on the onscreen keyboard that much easier. For those with dexterity issues, or simply for those of us with smaller hands using the larger iPhones, this option allows the onscreen keyboard to slide over to one side or the other to better facilitate typing.

1. Open any app that uses the onscreen keyboard.

   I'm using Notes for this example.

2. With the onscreen keyboard displayed, press-and-hold down on the emoji (smiley face) or international (globe) icon to display the Keyboard Settings menu, seen in **Figure 10-12**.

3. At the bottom of the Keyboard Settings menu, tap the left-sided keyboard icon or the right-sided keyboard icon to shift the keyboards keys in the desired direction.

4. Your keyboard will now appear either shifted to the left or right, as shown in **Figure 10-13**.

5. To quickly return to the standard keyboard, tap the white arrow to the left or right side of your shifted keyboard (its position depends on which side you shifted your keyboard).

Press and hold here

FIGURE 10-12

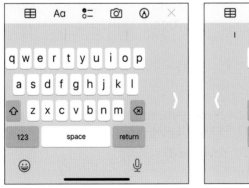 

FIGURE 10-13

Control Your iPhone with a
Tap on the Back

Your iPhone can perform a task or command by simply tapping on the back of the device! This feature is called Back Tap, and with a double- or triple-tap of your finger on the back of the iPhone, you can make the device do all kinds of cool things, such as

» Open an app.

» Mute your iPhone's speakers.

» Take a screenshot.

» Engage Siri.

» Initiate Accessibility functions.

» Launch shortcuts you've defined in the Shortcuts app.

To set up this handy feature:

1. Tap Settings and then tap Accessibility.

2. Scroll down and tap Touch, and then tap Back Tap at the bottom of the options list.

3. Tap either Double Tap or Triple Tap, depending on which action you want to assign a task to.

4. Select a task from the list of options in the Double Tap or Triple Tap screen.

5. To return to the Back Tap screen, tap Back Tap in the upper-left corner.

6. Double-tap or triple-tap the back of your iPhone any time you like and it will perform the assigned task.

Control Your iPhone with Voice Control

Another exciting accessibility feature is the ability to control your iPhone using your voice! The Voice Control feature also enables you to use numbers and grid overlays to command your iPhone to perform tasks. This feature is a real game-changer for a lot of folks.

1. Tap Settings and then tap Accessibility.

2. Scroll down and tap Voice Control. Do one of the following:

   - If you haven't used Voice Control before, on the screen that follows, tap Set Up Voice Control. Read through the information screens, tapping Continue to advance through them. At the end, you'll see the Voice Control toggle switch is set to on (green).

   Pay particular attention to the What Can I Say? screen. It tells you in simple terms the commands you can use to get started with Voice Control, such as "Go home" and "Show grid."

   - If you've used Voice Control, toggle the Voice Control switch on (green), as shown if **Figure 10-14**.

   You can easily tell when Voice Control is on: A blue circle containing a microphone appears in the upper-left corner of your iPhone's screen.

3. Tap Customize Commands to see what commands are built into Voice Control (see **Figure 10-15**), enable or disable commands, and even create your own commands. Tap Back in the upper-left to return to the main Voice Control options.

   I suggest taking your time in this section of the Voice Control options; you'll be surprised at what you can do out-of-the-gate with this amazing tool.

4. To add words to Voice Control (particularly useful with dictation, tap Vocabulary in the Voice Control options and then tap + in the upper-right. In the Add New Entry window, type the word or phrase, and then tap the Save button. Tap Voice Control in the upper-left to return to the main Voice Control screen.

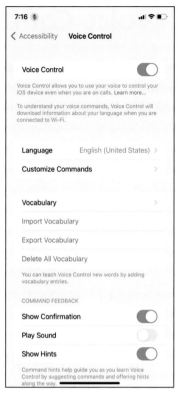

FIGURE 10-14

FIGURE 10-15

Overlays are a fantastic accessibility feature in iOS. If you use them, clickable items on the screen are labeled with numbers, names, or a numbered grid. Whenever you want to click an item, simply execute a command such as "tap three" to "tap" the item with your voice. Each of the three overlays are displayed in **Figure 10-16**.

You can enable an overlay in Settings ⇨ Accessibility ⇨ Voice Control ⇨ Overlay.

TIP

When you're not actively using the feature, the number and name labels will fade to a light gray so that you can more clearly see the screen. They darken again when you do use the feature.

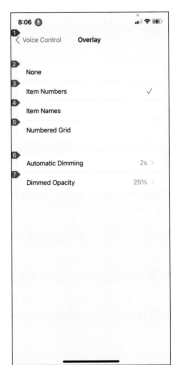
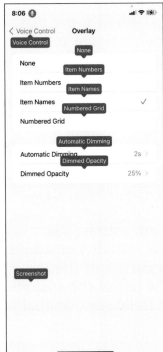
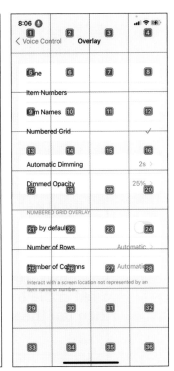

FIGURE 10-16

Adjust Accessibility Settings on a Per-App Basis

iOS 16's iteration of Accessibility features allows you to customize individually how each app handles display and text size settings, such as Bold Text, Larger Text, Increase Contrast, Smart Invert, and Reduce Motion. No need to settle for a one-setting-fits-all approach.

To customize Accessibility options on a per-app basis:

1. Tap Settings and then tap Accessibility.

2. Swipe down to the very bottom of the screen and tap Per-App Settings.

3. Tap Add App, and then tap an app in the list provided.

4. In the Per-App Settings screen that appears, tap the app you've just added, as shown in **Figure 10-17**.

5. On the screen that appears (see **Figure 10-18**), tap each option you'd like to customize, and then tap the name of the app in the upper-left corner to return to the list of options. Repeat this step until you're finished making customizations.

6. Tap Per-App Settings in the upper-left corner to return to the Per-App Settings screen.

7. If you want to customize additional apps, repeat Steps 3–6.

To delete a customization (don't worry, you're not deleting the app itself, just the Accessibility customizations you've made), tap Edit in the upper-right corner of the Per-App Settings screen, tap the – in a red circle to the left of the customization you want to remove, and then tap the red Delete button that appears on the right.

FIGURE 10-17

FIGURE 10-18

IN THIS CHAPTER

» **Activate Siri**

» **Translate, get suggestions, and call contacts**

» **Create reminders and alerts**

» **Add events to your calendar**

» **Play music, get directions, and ask for facts**

» **Search the web**

» **Send and dictate messages**

Chapter **11**

Conversing with Siri

One of the most talked about (pun intended) features on your iPhone is Siri, a personal assistant that responds to the commands you speak to your iPhone. With Siri, you can ask for nearby restaurants, and a list appears. You can dictate your email messages rather than type them. You can open apps with a voice command. Calling your mother is as simple as saying, "Call Mom." Want to know the capital of Rhode Island? Just ask. Siri checks several online sources to answer questions ranging from the result of a mathematical equation to the next scheduled flight to Rome (Italy or Georgia). You can have Siri search photos and videos and locate what you need by date, location, or album name. Ask Siri to remind you about an app you're working in, such as Safari, Mail, or Notes at a later time so you can pick up where you left off.

You can also have Siri perform tasks, such as returning calls and controlling the Music app. Siri can play music at your request and identify tagged songs (songs that contain embedded information that identifies them by categories such as artist or genre). You can also hail a ride with Uber or Lyft, watch live TV just by saying "Watch ESPN" (or say another app you might use, such as Netflix), find tagged photos, make payments with some third-party apps, and more. Siri can even hang up your calls and tell you what tasks you can accomplish with your apps.

Siri can offer you curated suggestions for Safari, Maps, and Podcasts. Siri also utilizes new voice technology that allows it to sound more natural and smooth, particularly when speaking long phrases. Siri can also process requests entirely on your iPhone, if it has an A12 Bionic chipset or newer (in other words, as long as it's a model from 2018 or later). This means requests you make of Siri are performed much faster than before. Whoop, whoop!

Activate Siri

When you first go through the process of registering your phone, you'll be prompted to begin making settings for your location, for using iCloud, and so on, and at one point you will see the option to activate Siri. As you begin to use your phone, iPhone reminds you about using Siri by displaying a message.

TIP

Siri requires internet access, and cellular data charges could apply when Siri checks online sources if you're not connected to Wi-Fi. In addition, available features may vary by area.

If you didn't activate Siri during the registration process, you can use Settings to turn on Siri by following these steps:

1. Tap the Settings icon on the Home screen.
2. Tap Siri & Search.
3. In the dialog in **Figure 11-1**, activate any or all of the following features by toggling the switch for the feature on (green):
   - *Listen for "Hey Siri":* Activate Siri for hands-free use. When you first enable "Hey Siri," you'll be prompted to set up the feature. Just walk through the steps to enable it and continue.

With this feature enabled, just say "Hey, Siri" and Siri opens, ready for a command. In addition, with streaming voice recognition, Siri displays in text what it's hearing as you speak, so you can verify that it has understood you correctly. This streaming feature makes the process of interacting with Siri faster.

REMEMBER

iPhone models older than the 6s and 6s Plus must be plugged into an outlet, a car, or a computer to use the "Hey Siri" feature.

- *Press Home/Side Button for Siri:* Activate Siri by pressing and holding down the Home button or the side button (if your iPhone doesn't have a Home button).

- *Allow Siri When Locked:* Use Siri even when the iPhone is locked.

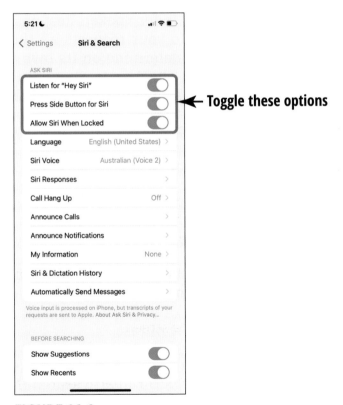

FIGURE 11-1

4. If you want to change the language Siri uses, tap Language and choose a different language in the list that appears.

5. To change the nationality or gender of Siri's voice from American to British or Australian (for examples), or from female to male, tap Siri Voice and make your selections. Some nationalities have multiple voices you can select.

TIP

Give several nationalities and genders a shot. Over the years, I've found that the Australian female voice (Voice 2) coupled with the English (United States) Language option was easier for me to understand, for whatever reason.

6. Let Siri know about your contact information by tapping My Information and selecting yourself in your contacts.

TIP

If you want to customize when Siri verbally responds to your requests, tap Siri Responses in the Siri & Search settings and choose from the selections. The Prefer Spoken Responses option causes Siri to verbally respond to your requests all the time — period. Automatic allows Siri to determine on its own when it's appropriate to verbally respond. The Always Show Siri Captions switch displays whatever Siri says on your iPhone's screen, while the Always Show Speech switch causes the transcript of your entire exchange with Siri to display.

Discover All That Siri Can Do

Siri allows you to interact by voice with many apps on your iPhone.

REMEMBER

No matter what kind of action you want to perform, first press and hold down on the Home button (or the side button for iPhone models without a Home button) until Siri opens. Or if you've enabled "Hey Siri," simply say the phrase.

You can pose questions or ask to do something like make a call or add an appointment to your calendar. Siri can also search the internet or use an informational service called Wolfram|Alpha to provide information on just about any topic.

Siri also checks with Wikipedia, Bing, and Twitter to get you the information you request. In addition, you can use Siri to tell iPhone to return a call, play your voicemail, open and search the App Store, control Music playback, dictate text messages, and much more.

Siri learns your daily habits and will offer suggestions throughout the day when appropriate. For example, say you usually stop by the local coffee shop around the same time each morning and use the shop's app to order a drink from your iPhone. Siri will pick up on this activity and eventually begin asking if you'd like to order a drink when you're within proximity of the coffee shop.

Siri is good at maintaining the context of questions. For example, if you ask something like "Where is the nearest Starbucks location?" and then follow it up with "What's the phone number there?" Siri will automatically know you're asking for the phone number of the Starbucks location it finds. It's not always perfect, but this feature does make for a vastly improved Siri experience.

Siri knows what app you're using, though you don't have to have that app open to make a request involving it. However, if you're in the Messages app, you can make a statement like "Tell Susan I'll be late," and Siri knows that you want to send a message. You can also ask Siri to remind you about what you're working on and Siri will note what you're working on and which app you're working in, and remind you about it at a later time you specify.

TIP

If you want to dictate text in apps like Notes or Mail, use the dictation icon on the onscreen keyboard to do so. See the task "Use Dictation," later in this chapter, for more about this feature.

Siri is able to announce possibly time-sensitive notifications when you're wearing AirPods (second generation or newer). For example, if you're on a run and it's getting close to time for a meeting, Siri will interrupt whatever you're listening to so that you're aware of the upcoming appointment.

Siri requires no preset structure for your questions; you can phrase things in several ways. For example, you might say, "Where am I?" to see a map of your current location, or you could say, "What is my current location?" or "What address is this?" and get the same results.

If you ask a question about, say, the weather, Siri responds both verbally and with text information (see **Figure 11-2**). Or Siri might open a form, as with email, or provide a graphic display for some items, such as maps. When a result appears, you can tap it to make a choice or open a related app.

FIGURE 11-2

Siri works with just about any app, including Phone, the App Store, Music, Messages, Reminders, Calendar, Maps, Mail, Weather, Stocks, Clock, Contacts, Notes, social media apps (such as Twitter), and Safari (see **Figure 11-3**). In the following tasks, I provide a quick guide to some of the most useful ways you can use Siri.

TIP

Siri supports many languages for translation, so you can finally show off those language lessons you took in high school. Some languages supported for translation include Chinese, Dutch, English, French, German, Italian, Spanish, Arabic, Danish, Finnish, Hebrew, Japanese, and Korean. However, Siri can speak to you in more languages when providing the results of inquiries. Visit www.apple.com/ios/feature-availability/#siri for an up-to-date list.

FIGURE 11-3

Get Suggestions

Siri anticipates your needs by making suggestions when you swipe from left to right on the initial Home screen page and tap in the search field at the top of the screen. Siri will list contacts you've communicated with recently, apps you've used, and nearby businesses, such as restaurants, gas stations, and coffee spots. If you tap an app in the suggestions, it will open and display the last viewed or listened to item.

Additionally, Siri lists news stories that may be of interest to you based on items you've viewed before.

Call Contacts

First, make sure that the person you want to call is entered in your Contacts app and include that person's phone number in their record. If you want to call somebody by stating your relationship, such as "Call sister," be sure to enter that relationship in the Add Related Name field in your sister's contact record. Also make sure that the settings for Siri (refer to Figure 11-1) include your own contact name in the My Information field. (See Chapter 7 for more about creating contact records.)

Follow these steps to call a contact:

1. Press and hold down on the Home button or side button (or say "Hey Siri," if you're using that feature) until Siri appears.

2. Speak a command, such as "Call Paul Whigham," "Return Popee's call," or "Call Mom." If you want to make a FaceTime call, you can say "FaceTime Mom."

3. If you have two contacts who might match a spoken name, tap one in the list that Siri provides, as shown in **Figure 11-4**, or state the correct contact's name to proceed.

The call is placed.

FIGURE 11-4

4. To end the call before it completes, press the Home button (side button for iPhone models without a Home button), or tap the end call button (white phone in the red circle).

To cancel any spoken request, you have three options: Say "Cancel," tap the Siri icon (swirling bands of light) on the Siri screen, or press and release the Home or side button (depending on your iPhone model). If you're using a headset or Bluetooth device, tap the end call icon on the device (refer to your device's documentation).

Create Reminders and Alerts

You can also use Siri with the Reminders app:

1. To create a reminder or alert, press and hold down on the Home button or the side button (if your iPhone doesn't have a Home button) and then speak a command, such as "Remind me to call Devyn tomorrow at 10 a.m." or "Wake me up Thursday at 6:15 a.m."

A preview of the reminder or alert is displayed, as shown in **Figure 11-5**.

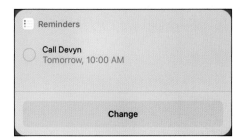

FIGURE 11-5

2. If you change your mind, tell Siri "Cancel" or "Remove."

3. If you want a reminder ahead of the event that you created, activate Siri and speak a command, such as "Remind me tonight about the play on Thursday at 8 p.m." A second reminder is created, which you can confirm or cancel if you change your mind.

Add Events to Your Calendar

You can also set up events on your Calendar using Siri. Press and hold down on the Home button or the side button (for iPhone models without a Home button) and then speak a phrase, such as "Set up a meeting for 1:30 p.m. tomorrow." Siri sets up the appointment (see **Figure 11-6**) and informs you that you can let it know if you'd like to make any changes.

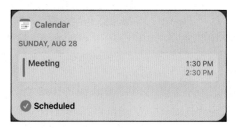

FIGURE 11-6

Play Music

You can use Siri to play music from the Music app.

1. Press and hold down on the Home button or the side button (iPhone models without a Home button) until Siri appears.

2. To play music, speak a command, such as "Play music" or "Play Jazz radio station" to play a specific song, album, or radio station, as shown in **Figure 11-7**.

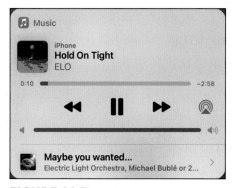

FIGURE 11-7

Apple has integrated Siri into Shazam, a music identifier app, to identify music. To use this integration:

1. When you're near an audio source playing music, press and hold down on the Home button or the side button (iPhone models without a Home button) to activate Siri.

2. Ask Siri a question, such as "What music is playing?" or "What's this song?"

3. Siri listens for a bit and, if Siri recognizes the song, it shows you the song name, artist, any other available information, and gives you the opportunity to purchase the music in the iTunes Store.

TIP

If you're listening to music or a podcast with earphones plugged in or connected wirelessly, and you stop midstream, the next time you plug in or wirelessly connect earphones, Siri recognizes that you might want to continue with the same item.

Get Directions

You can use the Maps app and Siri to find your current location, get directions, find nearby businesses (such as restaurants or a bank), or get a map of another location. Be sure to turn on Location Services to allow Siri to know your current location. Go to Settings and tap Privacy & Security ⇨ Location Services; make sure the Location Services toggle is switched on (green). Scroll farther down this same page, tap Siri & Dictation, tap While Using the App, and toggle the Precise Location switch on.

Here are some of the commands you can try to get directions or a list of nearby businesses:

» **"Where am I?"** displays a map of your current location.

» **"Where is Jordan-Hare Stadium?"** displays a map of that park's location, as shown in **Figure 11-8**.

» **"Find pizza restaurants."** displays a list of restaurants near your current location. Tap one to display a map of its location.

» **"Find PNC Bank."** displays a map with the location of the indicated business (or in some cases, several nearby locations, such as a bank branch and all ATMs).

» **"Get directions to the Empire State Building."** loads a map with a route drawn and provides a narration of directions to the site from your current location.

FIGURE 11-8

TIP

After a location is displayed on a map, tap the circled *i*-in a circle (information) button on the location's label to view its address, phone number, and website address, if available.

Ask for Facts

Siri uses numerous information sources, such as Wolfram|Alpha, Wikipedia, and Bing, to look up facts in response to questions. For example, you can ask, "What is the capital of Kansas?", "What is the square root of 2,300?", or "How large is Mars?" Just press and hold down on the Home button or the side button (if your iPhone doesn't have a Home button) and ask your question; Siri consults its resources and returns a set of relevant facts.

You can also get information about other things, such as the weather, stocks, or a scientific fact. Just say a phrase like one of these to get what you need:

» **"What is the weather?"** displays the weather report for your current location. If you want weather in another location, just specify the location in your question.

» **"What is the price of Apple stock?"** gets you the current price of the stock or the price of the stock when the stock market last closed.

» **"How hot is the sun?"** results in Siri telling you the temperature of the sun, in various unit conversions.

Search the Web

Although Siri can use its resources to respond to specific requests, such as "Who is the Queen of England?", more general requests for information will cause Siri to search further on the web. Siri can also search Twitter for comments related to your search.

For example, if you speak a phrase, such as "Find a website about birds" or "Find information about the World Series," Siri can respond in a couple of ways. The app can simply display a list of search results by using the default search engine specified in your settings for Safari. Or Siri can suggest, "If you like, I can search the web for such and such." In the first instance, just tap a result to go to that website. In the second, confirm that you want to search the web or cancel.

Send Email, Messages, or Tweets

You can create an email or an instant message by using Siri and existing contacts. For example, if you say, "Email Porter," a form opens already addressed to that stored contact. Siri asks for a subject and then a message. Speak your message contents and then say, "Send" to speed your message on its way.

Siri also works with messaging apps, such as Messages. If you have the Messages app open and you say, "Tell Keaton I'll call soon," Siri creates a message for you to approve and send.

Use Dictation

Text entry isn't Siri's strong point, but it's improving. Instead, you can use the dictation icon on the onscreen keyboard to speak text rather than type it. This feature is called Dictation.

1. Go to any app where you enter text, such as Notes or Mail, and tap in the document or form.

   The onscreen keyboard appears.

2. Tap the Dictation icon (microphone) in the lower right on the keyboard and speak your text.

3. To end the dictation, tap Done in the upper right.

TIP

When you finish speaking text, you can use the keyboard to make edits to the text Siri entered, although as voice recognition programs go, Dictation is pretty darn accurate. If a word sports a blue underline, which means there may be an error, you can tap to select the word and edit it.

3

Exploring the Internet and Apps

IN THIS PART . . .

Surfing the web

Sending and receiving email

Finding and managing apps

Participating in social media

IN THIS CHAPTER

» **Connect to the internet**

» **Navigate web pages and tabs**

» **Organize your searches with tab groups**

» **View history and search the web**

» **Use bookmarks**

» **Download files**

» **Translate web pages**

Chapter **12**

Browsing with Safari

After you're online, Safari, a built-in web browser (software that helps you navigate the internet), is your ticket to a massive world of information, entertainment, education, and more. Safari will look familiar to you if you've used a web browser on a PC or Mac computer, though the way you move around using the iPhone touchscreen may be new. If you've never used Safari, don't worry because I take you by the hand and show you all the basics of making it work for you.

In this chapter, you see how to go online with your iPhone, navigate among web pages, and use iCloud tabs to share your browsing experience between devices. Along the way, you place a bookmark for a favorite site, create a tab group, and then learn how to share said tab group, a feature new to Safari for iOS 16. You can also view your browsing history, edit bookmarks, save online images to the Photos app, search the web, or email or tweet a link to a friend. Finally, you explore how to translate web pages so you can read what's going on around the world, even if you may not speak the language.

Connect to the Internet

How you connect to the internet depends on which types of connections are available:

» You can connect to the internet via a Wi-Fi network. You set up this type of network in your home by using your computer and equipment from your internet provider. You can also connect over public Wi-Fi networks, or hotspots.

You might be surprised to discover how many hotspots your town or city has. Look for internet cafés, coffee shops, hotels, libraries, and transportation centers (such as airports or bus stations). Many of these businesses display signs alerting you to their free Wi-Fi.

» You can use the paid data network provided by AT&T, Sprint, T-Mobile, Verizon, or most any other cellular provider, to connect from just about anywhere you can get cellphone coverage through a cellular network.

To enable cellular data, tap Settings and then Cellular. Tap to toggle the Cellular Data switch on (green). Scroll further down the same screen to the list of apps, locate Safari in the list, and make sure the switch for it is toggled on (green).

WARNING

Browsing the internet using a cellular connection can eat up your data plan allotment quickly if your plan doesn't include unlimited data access. If you think you'll often use the internet with your iPhone away from a Wi-Fi connection, double-check your data allotment with your cellular provider or consider getting an unlimited data plan.

To connect to a Wi-Fi network, you have to complete a few steps:

1. Tap Settings on the Home screen and then tap Wi-Fi.

2. Be sure that Wi-Fi is on (green) and then choose a network to connect to by tapping it.

Network names should appear automatically when you're in their range. When you're in range of a public hotspot or if access to several nearby networks is available, you may see a message asking you to tap a network name to select it. After you select a network you may see a message asking for your password. Ask the owner of the hotspot (for example, a hotel desk clerk or business owner) for this password, or enter your own network password if you're connecting to your home network.

Free public Wi-Fi networks usually don't require passwords, or the password is posted prominently for all to see. (If you can't find the password, don't be shy about asking someone.)

3. Tap the Join button when prompted.

You're connected! Your iPhone will now recognize the network and you can connect to it again later without entering the password.

After you connect to public Wi-Fi, someone else could possibly track your online activities because these are unsecured networks. When connected to a public hotspot, avoid accessing financial accounts, making online purchases, or sending emails containing sensitive information.

Explore Safari

Safari is iPhone's default web browsing app.

This is not the Safari of old. Safari changed a lot with iOS 15 — and the changes continue in iOS 16. So even if you're familiar with previous versions of iOS, it's worthwhile to peruse this chapter.

You can change your iPhone's default browser from Safari to another you've downloaded. Go to Settings, find and tap the name of the browser you want to use, tap Default Browser App, and then tap the name of the browser you want to use as default.

1. After you're connected to a network, tap the Safari icon, which is on the dock at the bottom of the Home screen.

Safari opens, possibly displaying the Apple iPhone Home page the first time you go online (see **Figure 12-1**).

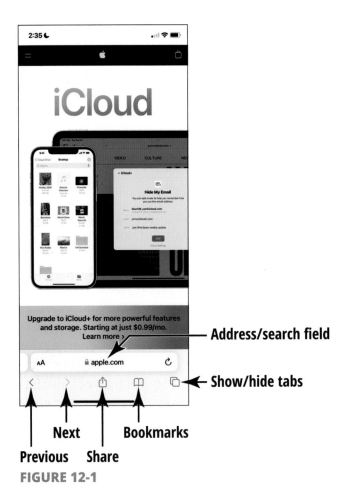

Address/search field

Show/hide tabs

Next   Bookmarks

Previous   Share

FIGURE 12-1

2. Hold down two fingers (thumb and forefinger may be easiest) together on the screen and spread them apart to expand the view (also known as zooming in). Hold down your fingers on the screen about an inch or so apart and quickly bring them together to zoom back out.

You can also double-tap the screen with a single finger to restore the default view size. (If you tap a link, though, your gesture will just open that link.)

TIP

Using your fingers on the screen to enlarge or reduce the size of a web page allows you to view the screen at various sizes, giving you more flexibility than the double-tap method.

3. Put your finger on the screen and flick upward to scroll down on the page.

4. To return to the top of the web page, put your finger on the screen and drag downward or tap the status bar at the very top of the screen.

Navigate Web Pages

Here are some basic steps to give you a feel for getting around various websites:

1. Tap in the address field near the bottom of the screen. The onscreen keyboard appears, as shown in **Figure 12-2**.

2. Enter a web address, such as www.dummies.com.

TIP

By default, autofill is turned on in iPhone, so entries you make in fields, such as the address and password fields, automatically display possible matching entries. You can turn off autofill for your contact and credit card information by going to Settings ⇨ Safari ⇨ AutoFill, and toggling the switch off for one or both items. You can turn off autofill for passwords by going to Settings ⇨ Passwords ⇨ Password Options and toggling the AutoFill Passwords switch off.

3. Tap the Go key on the keyboard to display the website.

- If a page isn't displayed properly, tap the reload icon (circular arrow), at the right end of the address field.

- If Safari is loading a web page and you change your mind about viewing the page, you can stop loading the page. Just tap the cancel icon (X), at the right end of the address field.

4. Tap the previous icon (<) at the bottom of the screen to go to the last page you displayed.

5. Tap the next icon (>) at the bottom of the screen to go forward to the page you just backed up from.

6. To follow a link to another web page (links are typically indicated by colored text or graphics), tap the link with your finger.

To view the destination web address of the link before you tap it, just touch and hold down on the link. A menu appears, along with a preview of the site, as shown in **Figure 12-3**. Tap Hide Preview in the upper right of the preview window to see the full link address.

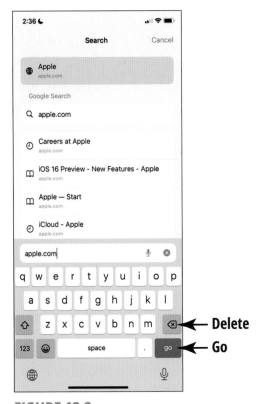

FIGURE 12-2

FIGURE 12-3

Apple QuickType supports predictive text in the onscreen keyboard. This feature adds the capability for iPhone to spot what you probably intend to type from text you've already entered and suggest it to save you time typing.

Use Tabbed Browsing

Tabbed browsing enables you to have several websites open at one time so that you can move easily among those sites.

1. With Safari open and a web page already displayed, tap the show tabs icon in the bottom-right corner (refer to Figure 12-1).

   The new tab view appears.

   TIP If you don't see the show tabs icon at the bottom of the screen, just swipe down on the screen and it will reappear.

2. To add a new page (meaning that you're opening a new website), tap the new tab icon (+) in the lower left of your iPhone's screen.

   A page with your favorite sites and an address field appears, like the one in **Figure 12-4**.

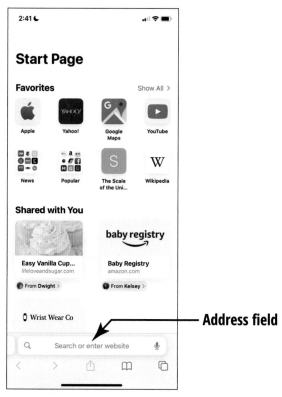

Address field

FIGURE 12-4

TIP

You can get to the same new page by simply tapping in the address field on any site.

3. Tap in the address field and use the onscreen keyboard to enter the web address for the website you want to open. Then tap the Go key, and the website opens on the page.

TIP

4. Repeat Steps 1 to 3 to open as many new web pages as you'd like.

5. Switch among open sites by tapping outside the keyboard to close it and then tapping the show tabs icon and scrolling among recent sites by swiping up or down. Find the one you want and then tap to open it.

TIP

You can easily rearrange sites in the tabs window. Just tap and hold down on the tab you want to move and drag it up or down the list until it's in the spot you want. (The other sites in the window politely move to make room.) To drop it in the new location, simply remove your finger from the screen.

6. To delete a tab, tap the show tabs icon, scroll to locate the tab, and then tap the close icon (X) in the upper-right corner of the tab.

The close icon may be difficult to see on some sites, but trust me, it's there.

Organize with Tab Groups

The tab groups feature in iOS 16 enables you to keep similar tabs together so that they're easier to organize and find. This feature is especially helpful if you're someone who likes to keep a million tabs open at once; tab groups keep you from having to swipe until your fingers bleed to find the site tab you're looking for. A new function in iOS 16 is the capability to share tab groups. Anyone participating in the shared tab group can add or delete tabs in the group; others in the group see the changes instantly.

1. With Safari open, tap the show tabs icon in the bottom-right corner (refer to Figure 12-1).

2. In the lower-middle of the screen, tap *x* Tabs, where *x* represents the number of open tabs.

   Figure 12-5 shows that I have 14 open tabs in Safari.

3. In the Tab Groups window that appears (see **Figure 12-6**), you can do any of the following:

   - Tap a group to open it.

   - Tap New Empty Tab Group to create a new group. You'll be prompted to give it a descriptive name (such as Food in Figure 12-6).

   - Tap New Tab Group from *x* Tabs to create a tab group from the tabs you currently have open. Again, give the new group a descriptive name to help you stay organized.

   - Tap Done when you're finished.

FIGURE 12-5

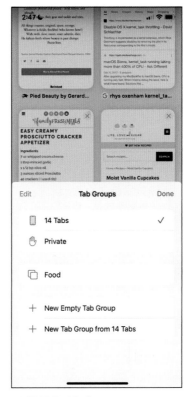

FIGURE 12-6

4. To move a tab to a tab group:

   (a) Tap and hold down on the address field at the bottom of the screen until a menu opens.

   (b) Tap the Move to Tab Group option.

   (c) Tap the name of the group you want to move the tab to (see **Figure 12-7**) and it will join that tab group (see **Figure 12-8**).

5. To share a tab group:

   (a) Scroll to the top of your tab group until you see its name.

   (b) Tap the share icon in the top-right corner (labeled in Figure 12-8).

FIGURE 12-7

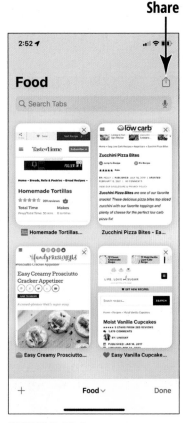

FIGURE 12-8

(c) Tap the name of the person you want to share the group with from the list of suggested contacts that appears, or tap the Messages icon to select other recipients.

(d) Tap the blue arrow icon to send the link to your shared tab group to the folks you want to share it with. After your friends click the link, effectively accepting your invitation to the group, they can participate in your shared tab group.

View Browsing History

As you move around the web, your browser keeps a record of your browsing history. This record can be handy when you want to visit a site you viewed previously but whose address you've now forgotten, or if you accidentally close a site and want to quickly reopen it.

1. With Safari open, tap the bookmarks icon (refer to Figure 12-1).

TIP

After you master the use of the bookmarks icon options, you might prefer a shortcut to view your browsing history. Tap and hold down on the previous icon (<) at the bottom left on any screen, and your browsing history for the current tab appears. You can also tap and hold down on the next icon (>) to look at pages you backtracked from.

2. On the menu shown in **Figure 12-9**, tap the History tab (clock).

3. In the history list that appears (see **Figure 12-10**), tap a site to navigate to it. Tap Done to leave History and return to browsing.

TIP

To clear your history, tap the Clear button (refer to Figure 12-10). On the screen that appears, tap The Last Hour, Today, Today and Yesterday, or All Time. The Clear button is useful when you don't want your spouse or grandchildren to see where you've been browsing for anniversary, birthday, or holiday presents!

Bookmarks tab   **History tab**

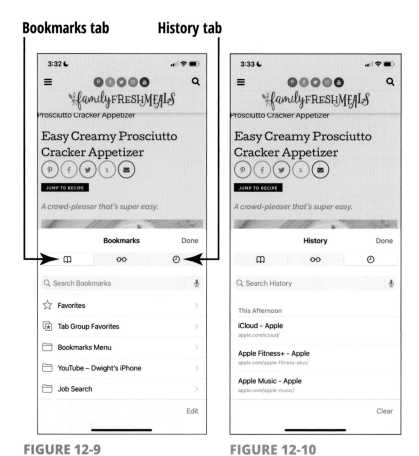

FIGURE 12-9   **FIGURE 12-10**

Search the Web

If you don't know the address of the site you want to visit (or you want to research a topic or find other information online), get acquainted with Safari's search feature on iPhone. By default, Safari uses the Google search engine.

1. With Safari open, tap the address field (refer to Figure 12-1). The onscreen keyboard appears.

TIP

To change your default search engine from Google to Yahoo!, Bing, DuckDuckGo, or Ecosia, tap Settings, tap Safari, and then tap Search Engine. Tap Yahoo!, Bing, DuckDuckGo, or Ecosia, and your default search engine changes.

2. Begin entering a search term.

   With recent versions of Safari, the search term can be a topic or a web address because of what's called the unified smart search field.

3. Tap one of the suggested sites or complete your entry and tap the Go key (see **Figure 12-11**) on your keyboard.

4. In the search results that are displayed, tap a link to visit that site.

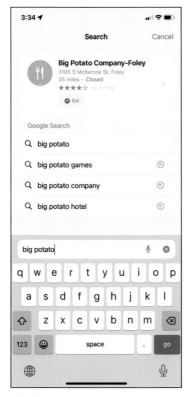

FIGURE 12-11

Add and Use Bookmarks

Bookmarks are a way to save favorite pages so that you can easily visit them again.

1. With a page open that you want to bookmark, tap the share icon (upward-pointing arrow in a box) at the bottom of the screen.

TIP

If you want to sync your bookmarks on your iPhone browser, go to Settings on iPhone and make sure that iCloud is set to sync with Safari. To do so, go to Settings ⇨ Apple ID ⇨ iCloud ⇨ Show All, and make sure the Safari switch is set on (green).

2. On the menu that appears (see **Figure 12-12**), tap Add Bookmark. (You may need to swipe up the page to see it.)

The Add Bookmark screen appears, as shown in **Figure 12-13**.

3. You can edit the bookmark name by tapping it and using the onscreen keyboard.

FIGURE 12-12

FIGURE 12-13

4. Tap the Save button in the upper-right corner.

   The item is saved to your favorites by default; to save to another location, tap Favorites (or whatever is listed in the Location section of the Add Bookmark dialog) and choose the desired location from the list.

5. To go to the bookmark, tap the Bookmarks tab (refer to Figure 12-9). On the Bookmarks menu that appears, tap the bookmarked page that you want to visit. If you saved the bookmark to a folder, tap the folder first to open it.

When you tap the bookmarks icon, you can tap Edit in the lower-right corner and then use the New Folder option (in the lower-left corner) to create folders to organize your bookmarks or folders. When you next add a bookmark, you can then choose, from the dialog that appears, any folder to which you want to add the new bookmark.

You can reorder your bookmarks easily. Tap the bookmarks icon, tap the Edit button, and find the bookmark you want to rearrange. Tap and hold down on the three parallel lines to the right of the bookmark, and then drag it up or down the list (other bookmarks will kindly move out of the way as you go), releasing it once you get to the place you'd like it to reside. You can also delete bookmarks from the same screen by tapping the red circle to the left of a bookmark and then tapping the red Delete button that appears to the right.

Download Files

Download Manager for Safari helps you efficiently download files from websites and store them to a location of your choosing. You can opt to store downloaded files on your iPhone or in iCloud.

Set the default download location for files you download in Safari. Go to Settings ⇨ Safari ⇨ Downloads and tap the location you want to use.

1. Open a site in Safari that contains a file you'd like to download.

2. Tap and hold down on the link for the file until the menu shown in **Figure 12-14** appears.

3. Tap the Download Linked File option to download the file to your iPhone, to iCloud, or to another destination.

The downloads icon (downward-pointing arrow) appears to the left of the address field at the bottom of the screen, as shown in **Figure 12-15**.

4. Tap the downloads icon to see the progress of the download.

5. When the download is finished, tap it in the Download Manager menu to open it, or tap the magnifying glass to see where the file is stored.

FIGURE 12-14

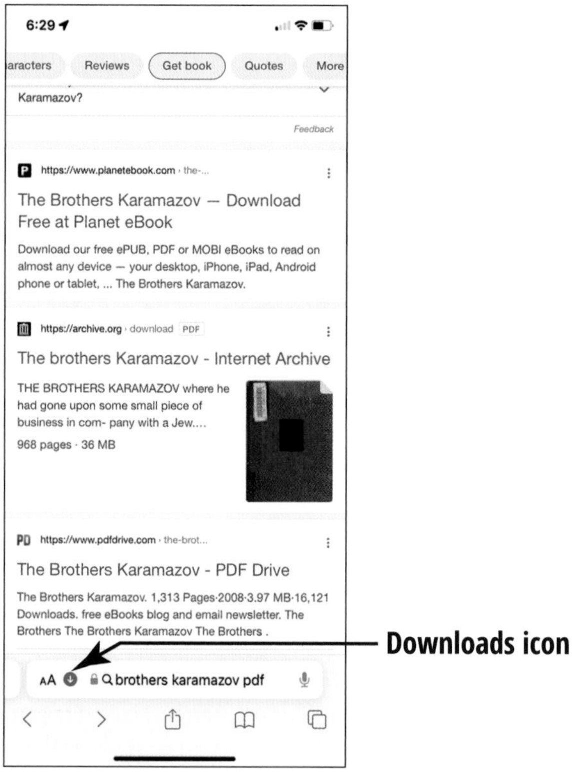

FIGURE 12-15

Downloads icon

Translate Web Pages

Safari includes a great trick: web page translation! Visit a compatible web page and Safari can translate it into many languages (with more sure to come): English, Spanish, Brazilian Portuguese, Simplified Chinese, German, Russian, Turkish, Thai, Polish, Vietnamese, Dutch, Indonesian, and French.

1. Open a site in Safari that's in a language you'd like to translate.

2. Tap ᴀA in the URL field (see **Figure 12-16**), and then tap the Translate To option. If you don't see this option, Safari is unable to translate the site.

The page is translated into the language you selected.

Tap here to translate page

FIGURE 12-16

IN THIS CHAPTER

» **Add an email account**

» **Read, reply to, or forward email**

» **Create, format, and send email**

» **Search email**

» **Mark or flag email**

» **Create events with email contents**

» **Delete and organize email**

» **Create a VIP list**

Chapter **13**

Working with Email in the Mail App

S taying in touch with others by email is a great way to use your iPhone. You can access an existing account using the handy built-in Mail app on your iPhone or sign in to your email account using the Safari browser. In this chapter, we take a look at using Mail, which involves adding an existing email account. Then you can use Mail to write, format, retrieve, and forward messages from that account, right from your iPhone.

Mail offers the capability to mark the messages you've read, delete messages, and organize your messages in folders, as well as use a handy search feature. You can create a VIP list so that you're notified when that special person sends you an email.

In this chapter, you familiarize yourself with the Mail app and its various features.

Add an Email Account

You can add one or more email accounts, including the email account associated with your iCloud account, using your iPhone's Settings app. If you have an iCloud, Microsoft Exchange (often used for business accounts), Gmail, Yahoo!, AOL, or Outlook.com (this includes Microsoft accounts from Live, Hotmail, and so on) account, iPhone pretty much automates the setup.

TIP

If you have an iCloud account and have signed in to it already, your iCloud email account will already be set up for you in Mail.

TIP

If this is the first time you're adding an account, and if you need to add only one, save yourself a few taps; just open Mail and begin from Step 4.

1. To set up iPhone to retrieve messages from your email account at one of these popular providers, first tap the Settings icon on the Home screen.

2. In Settings, tap Mail, and then Accounts. The screen shown in **Figure 13-1** appears.

3. Tap Add Account and the options shown in **Figure 13-2** appear.

4. Tap iCloud, Microsoft Exchange, Google, Yahoo!, AOL, or Outlook. com. Enter your account information in the form that appears and follow any instructions to complete the process.

   Each service is slightly different, but none are complicated. If you have a different email service than these, skip to the next section, "Manually Set Up an Email Account."

Tap here

FIGURE 13-1   **FIGURE 13-2**

5. After your iPhone takes a moment to verify your account information, on the next screen (see **Figure 13-3**), you can tap any on/off switch to have services from that account synced with your iPhone.

6. When you're done, tap Save in the upper-right corner of the screen.

   The account is saved, and you can now open it using Mail.

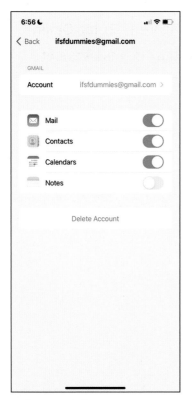

FIGURE 13-3

Manually Set Up an Email Account

You can also set up many email accounts, such as those available through Earthlink or a cable provider's service, by obtaining the host name from the provider. To set up an existing account with a provider other than iCloud (Apple), Microsoft Exchange, Gmail (Google), Yahoo!, AOL, or Outlook.com, you enter the account settings yourself.

TIP

If this is the first time you're adding an account and you need to add only one, just open Mail and begin from Step 3.

1. Tap the Settings icon on the Home screen.

2. In Settings, tap Mail, tap Accounts, and then tap the Add Account button (refer to Figure 13-1).

3. On the screen that appears (refer to Figure 13-2), tap Other.

4. On the screen shown in **Figure 13-4**, tap Add Mail Account.

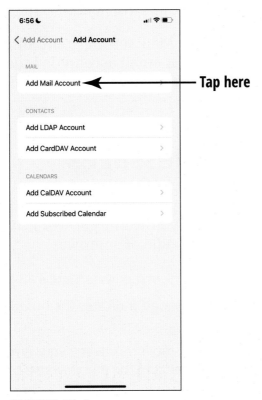

Tap here

FIGURE 13-4

5. In the form that appears, enter your name and an account email address, a password, and a description, and then tap Next.

iPhone takes a moment to verify your account.

TIP

iPhone will probably add the outgoing mail server (SMTP) information for you. If it doesn't, you may have to enter it yourself. If you have a less mainstream email service, you may have to enter the mail server protocol (POP3 or IMAP — ask your provider for this information) and your password.

6. On the next screen, tap either IMAP or POP, enter the appropriate information in the required fields, and then tap Save to return to the Accounts screen.

You can now access the account through iPhone's Mail app.

If you turn on Calendars in the Mail settings, any information you've put in your calendar in that email account is brought over to the Calendar app on your iPhone (discussed in more detail in Chapter 22).

Open Mail and Read Messages

Now for the exciting part: opening and reading your email! It's kinda like checking your mailbox, but you won't get wet if it's raining or eaten up by mosquitoes in the middle of summer.

1. Tap the Mail app icon on the Home screen (see **Figure 13-5**).

A red circle on the icon, called a badge, indicates the number of unread emails in your inbox.

2. In the Mail app (see **Figure 13-6**), tap Inbox to see your emails. If you have more than one account listed, tap the inbox whose contents you want to display.

3. To read a message, simply tap it to open it.

The 3D Touch feature allows you to preview an email before you open it. Simply press lightly and hold down on an email in the inbox to open a preview of the message. From the preview, you can elect to perform several functions, such as Reply, Forward, Mark as Read or Unread, Archive, or send to Trash. If you want to view the entire message, release the hold and tap the preview.

4. If you need to scroll to see the entire message, flick upward to scroll down.

Tap the Mail icon

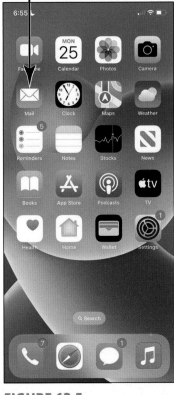

FIGURE 13-5

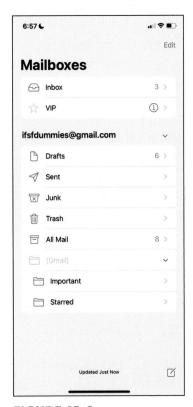

FIGURE 13-6

TIP

Tap the next or previous icons (top-right corner of the message) to move to the next or previous message in the inbox, or tap the back icon (<) in the top left to return to your inbox.

Email messages that you haven't read are marked with a blue circle in your inbox. After you read a message, the blue circle disappears. You can mark a read message as unread to help remind you to read it again later. With the inbox displayed, swipe to the right on a message (starting your swipe just a little in from the left edge of the screen), and then tap Unread. If you swipe quickly and all the way to the right, you don't need to tap; the message will be marked as unread automatically.

Reply To or Forward Email

Replying to email is just like replying to snail mail; it's the nice thing to do. Since we're all nice people, let's find out how to reply to those good folks who are sending us messages.

1. With an email message open, tap the reply icon (left-facing arrow) at the bottom of the screen. Then tap Reply, Reply All (available if there are multiple recipients), or Forward in the menu that appears (see **Figure 13-7**).

FIGURE 13-7

2. In the new email message that appears (see **Figure 13-8**), tap in the To field and enter another addressee if you like (you have to do this if you're forwarding). Next, tap in the message body and enter a message (see **Figure 13-9**).

Tap to add a recipient

Tap in the message body to add content

FIGURE 13-8

FIGURE 13-9

TIP

If you want to move an email address from the To field to the Cc or Bcc field, press and hold down on the address and drag it to the other field.

3. Tap the Send button in the upper-right corner (blue circle containing an upward-pointing arrow). The email goes on its way.

TIP

If you tap Forward to send the message to someone else and the original message had an attachment, you're offered the option of including or omitting the attachment.

Create and Send a New Message

It's time to reach out to everyone to plan for the family reunion, or maybe you need to provide vacation information to a friend who's accompanying you. "How to create and send that new email?" you ask? Well, here's how:

1. With Mail open, tap the compose icon (paper and pencil) in the bottom-right corner. A blank email appears (see **Figure 13-10**).

FIGURE 13-10

2. Enter a recipient's address in the To field by tapping the field and typing the address. If you have addresses in Contacts, tap the plus sign (+) in the To field to choose an addressee from the Contacts list that appears.

3. If you want to send a copy of the message to other people, tap the Cc/Bcc field. When the Cc and Bcc fields open, enter addresses in either or both.

   Use the Bcc field to specify recipients of blind carbon copies, which means that no other recipients are aware that that person received this reply.

4. In the Subject field, enter the subject of the message.

5. Tap in the message body and type your message.

6. If you want to check a fact or copy and paste some part of another message into your draft message, swipe down near the top of the email to display your inbox and other folders. Locate the message, and when you're ready to return to your draft, tap the subject of the email, which is displayed near the bottom of the screen.

7. When you've finished creating your message, tap the send icon (upward-pointing arrow in a blue circle) in the upper-right corner.

You can also schedule your email to send later. Instead of tapping the send icon, hold down on the send icon until a menu appears. Select Send Now, Send 9:00 PM Tonight, Send 8:00 AM Tomorrow, or Send Later (customize the date and time to send your email).

TIP

Oops! Did you forgot to mention something or accidentally send the email to the wrong recipient? Well, if you're quick about it, iOS 16 lets you undo your action. Immediately after you send an email, you'll see an Undo Send button at the very bottom of the screen. Tap that button within 10 seconds to open the draft of your previous email message, and then tap the Cancel button in the upper-left corner to save the email to your Drafts folder. Remember: You have only 10 seconds after tapping the Send button to undo it!

Format Email

You can apply some basic formatting to email text. You can use bold, underline, and italic formats, and indent text using the Quote Level feature.

1. Press and hold down on the text in the message you're creating and choose Select or Select All to select a single word or all the words in the email (see **Figure 13-11**).

If you need to select a single word, make sure to press and hold down on that word, rather than on just any old word in the text.

TIP

When you make a selection, blue handles appear that you can drag to add adjacent words to your selection. If the toolbar disappears after you select the text, just tap one of the selection handles and the toolbar will reappear.

2. To apply bold, italic, or underline formatting, tap the Format button on the toolbar that appears when you press and hold down on the text. To see more tools, tap the right or left arrows on the toolbar.

3. In the toolbar that appears (see **Figure 13-12**), tap Bold, Italic, or Underline to apply formatting.

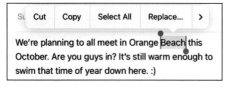

FIGURE 13-11   FIGURE 13-12

4. To change the indent level, press and hold down on the beginning of a line and then tap Quote Level in the toolbar that appears. (You might have to tap an arrow on the right or left of the toolbar to see it.) Tap Increase to indent the text or Decrease to move indented text farther toward the left margin.

Mail in iOS 16 allows you to go beyond the basics, though. It includes much-improved text formatting and font support, freeing you up to create some great looking emails. However, it doesn't stop there: The format bar (which appears above the keyboard, as shown in **Figure 13-13**) allows you to easily jazz up your email with a variety of options.

FIGURE 13-13

TIP

If you see words above the keyboard and not the format bar, tap the arrow to the right of the words to bring the format bar back into view.

Here's a quick look at the options in the format bar, from left to right:

» **Desktop-class text formatting:** Tap the Aa icon to see a bevy of formatting options (shown in **Figure 13-14**), such as

• Choose Bold, Italic, Underline, and Strikethrough. (Okay, these options aren't new, but the rest are.)

• Change the font by tapping Default Font and browsing a surprisingly extensive list of fonts to choose from.

- Decrease or increase the text size by tapping the small *A* or the large *A*, respectively.
- Tap the color wheel to select a color for your text.
- Insert numbered or bulleted lists.
- Select left, center, or right justification.
- Increase or decrease the quote level.
- Indent or outdent paragraphs.

» **Insert photo or video:** Tap to insert a photo or video from the Photos app.

» **Camera:** Tap to insert a new photo or video directly from the Camera app.

FIGURE 13-14

ALTERNATIVE EMAIL APPS

You can choose from lots of great email apps for your iPhone if Mail isn't what you're used to (or if you simply don't like it). Here are a few of the better options: Gmail (the official app for Google Mail), Outlook (Microsoft's official app for Outlook), Edison Mail, Airmail, Spark, and Yahoo! Mail (Yahoo!'s official app).

iOS 16 offers a way for you to replace Mail with a third-party email app as your default. Just go to Settings, scroll down until you see the name of your favorite email app, and tap it. Then tap the Default Email App option to set it as default.

» **Scan document:** Tap to scan a paper document and add it to your email.

» **Live text:** Tap to scan a paper document or image for text, which is then inserted into your email.

» **Attachment:** Tap to add an attachment from the Files app to the email. (See Chapter 4 for more info about Files.)

» **Insert drawing:** Tap to create a drawing and insert it in your email.

Search Email

What do you do if you want to find all messages from a certain person or containing a certain word? You can use Mail's handy Search feature to find these emails.

1. With Mail open, tap an account to display its inbox.

2. In the inbox, tap and drag down near the top email to display the search field. Tap in the search field, and the onscreen keyboard appears.

You can also use the Search feature covered in Chapter 2 to search for terms in the To, From, or Subject lines of mail messages.

TIP

3. Tap the All Mailboxes tab to view messages that contain the search term in any mailbox, or tap the Current Mailbox tab to see only matches in the current mailbox.

These options may vary slightly depending on which email service you use.

4. Enter a search term or name, as shown in **Figure 13-15**. If multiple types of information are found, such as People or Subjects, tap the one you're looking for.

Matching emails are listed in the results.

FIGURE 13-15

TIP

To start a new search, tap the delete icon (X in a gray circle) at the end of the search field to delete the term and start over. To go back to the full inbox, tap Cancel to end your search.

Mark Email as Unread or Flag for Follow-Up

You can use a simple swipe to access tools that either mark an email as unread after you've read it (placing a blue dot to the left of the message) or flag an email (which by default places an orange flag to the right of it, although you can choose an alternate color). If the email is both marked as unread and is flagged, both a blue dot and an orange flag will appear on the message. These methods help you to remember to reread an email that you've already read or to follow up on a message at a later time.

1. With Mail open and an inbox displayed, swipe to the left on an email to display three options: More, Move, and Trash (or Archive).

TECHNICAL STUFF

    Whether you see Archive or Trash depends on the type of email account you're using and how it has been set up by the email provider.

2. Tap More.

    On the menu, you're given several options, including Mark as Read or Mark as Unread (depending on the current state of the email) and Flag. To see the full list of options, swipe up (see **Figure 13-16**).

TIP

    You can get to the Mark as Read or Mark as Unread command also by swiping from left to right on a message displayed in your inbox.

3. To mark a message as read or unread, tap the appropriate command.

    You return to your inbox.

4. To assign a flag to the email, tap Flag and choose a color for the flag (again, the default is orange). To remove the flag, just tap More and then tap Unflag.

TIP

On the menu shown in Figure 13-16, you can also select the Mute or Notify Me commands. Mute allows you to mute a thread of emails that just won't stop bugging you. Notify Me causes Mail to notify you when someone replies to this email thread.

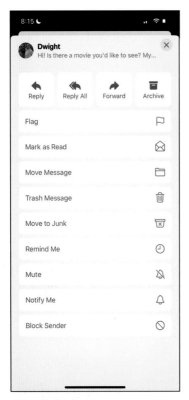

FIGURE 13-16

Create an Event from Email Contents

A neat feature in Mail is the ability to create a calendar event from an email. To test this out:

1. Create an email to yourself mentioning a reservation on a specific airline on a specific date and time.

   You could instead mention another type of reservation, such as for dinner, or mention a phone number.

2. Send the message to yourself and then open Mail.

3. In your inbox, open the email.

   The pertinent information is displayed in underlined text.

4. Tap the underlined text, and the menu shown in **Figure 13-17** appears.

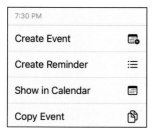

7:30 PM

Create Event

Create Reminder

Show in Calendar

Copy Event

FIGURE 13-17

5. Tap Create Event to display the New Event form from Calendar.

6. Enter additional information about the event, and then tap Done.

TIP

Siri may also detect an event in your email. If so, you'll see a notification at the top of the email that Siri did indeed find an event. Tap the small Add button to quickly create the event.

Delete Email

When you no longer want an email cluttering your inbox, you can delete it.

1. With the inbox displayed, tap Edit in the upper-right corner. A circular button is displayed to the left of each message (see **Figure 13-18**).

2. Tap the circle next to the message(s) you want to delete. A message marked for deletion has a check mark and the circle turns blue.

You can tap multiple items if you have several emails to delete.

TIP

3. Tap the Trash or Archive button at the bottom right of the screen. The selected messages are moved to the Trash or Archive folder.

What's the difference between the Trash and Archive folders? Basically, email sent to a Trash folder typically is deleted forever after a certain amount of time (usually 30 days). Email sent to an Archive folder is removed from the inbox but kept indefinitely for future use.

TECHNICAL STUFF

FIGURE 13-18

Tap to select a message

Selected message

TIP

You can delete an open email also by tapping the trash or archive icon on the toolbar at the bottom of the screen, or by swiping from right to left on a message displayed in an inbox and tapping the Trash or Archive button that appears.

Organize Email

You can move messages into any of several predefined folders in Mail, or you can create your own. (The predefined folders vary depending on your email provider and the folders you've created on your provider's server.)

1. After displaying the folder containing the message you want to move (for example, the Inbox folder), tap the Edit button.

Circular buttons are displayed to the left of each message (refer to Figure 13-18).

2. Tap the circle next to the message you want to move. Select multiple messages if you like.

3. Tap Move at the bottom of the screen.

4. In the Mailboxes list that appears (see **Figure 13-19**), tap the folder where you want to store the message.

The message is moved.

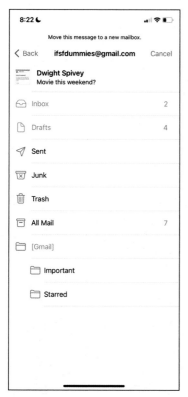

FIGURE 13-19

Create a VIP List

A VIP list is a way to create a list of senders that you deem more important than others. When any of these senders sends you an email, you'll be notified of it through the Notifications feature of iPhone.

1. In the main list of Mailboxes, tap the info icon (circled *i*) next to the VIP option (see **Figure 13-20**).

2. In the VIP List screen that appears, tap Add VIP, and your Contacts list appears.

3. Tap a contact to make that person a VIP.

4. To make settings for whether VIP mail is flagged in Notification Center, press the Home button or swipe up from the bottom of the screen (if your iPhone uses Face ID), and then tap Settings.

5. Tap Notifications, and then tap Mail. In the settings that appear, shown in **Figure 13-21**, tap Customize Notifications.

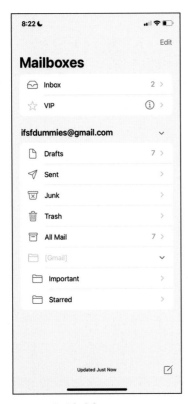

FIGURE 13-20

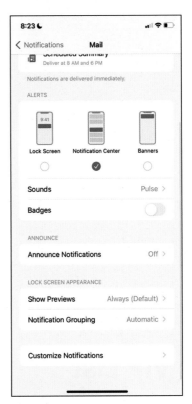

FIGURE 13-21

6. Tap VIP, and then toggle the Alerts switch to on (green).

7. To close Settings, press the Home button or swipe up from the bottom of the screen (on iPhone models without a Home button).

New mail from your VIPs should now appear in Notification Center when you swipe down from the top of the screen. And, depending on the settings you chose, new mail may cause a sound to play, a badge icon to appear on your lock screen, or a blue star icon to appear to the left of the message in the inbox in Mail. Definitely VIP treatment.

IN THIS CHAPTER

» **Explore and get senior-recommended apps in the App Store**

» **Organize your apps on Home screens and in folders**

» **Delete apps you no longer need**

» **Update apps**

» **Purchase, download, and subscribe to games**

» **Challenge friends in Game Center**

Chapter **14**

Expanding Your iPhone Horizons with Apps

Some apps (short for applications), such as Health and Maps, come preinstalled on your iPhone. But you can choose from a world of other apps out there for your iPhone, some for free (such as Facebook or Instagram) and some for a price (typically 99 cents to about $10, though some can top out at much steeper prices).

Apps range from games to financial tools (such as loan calculators) to apps that help you when you're planning an exercise regimen or going grocery shopping. Still more apps are developed for use by private entities, such as hospitals, businesses, and government agencies.

In this chapter, I suggest some apps that you might want to check out, explain how to use the App Store feature of your iPhone to find, purchase, and download apps, and detail how to organize your apps. You also find out a bit about having fun with games on your iPhone.

Explore Senior-Recommended Apps

As I write this book, new iPhone apps are in development, so even more apps that could fit your wants and needs are available seemingly every day. To get you exploring what's possible, I provide a quick list of apps that might whet your appetite.

Access the App Store by tapping the App Store icon on the Home screen. You can start by exploring the Today tab (which features special apps and articles), by checking out categories in the Apps tab, by searching for games in the Games tab, or by subscribing to Apple Arcade games (see the buttons along the bottom of the screen). Or you can tap Search and find apps on your own. Tap an app to see more information about it.

Here are some interesting apps to explore:

>> **April Coloring — Oil Painting by Number (free, with in-app purchases available):** Adult coloring books have exploded in popularity over the years, and they're a wonder at helping us unwind and unleash our creativity. This app, shown in **Figure 14-1,** is one of the best coloring book apps for iPhone, emulating oil painting to near perfection.

>> **Blood Pressure Monitor (free, with in-app purchases available):** This app helps you keep track of your blood pressure and maintain records over extended periods of time in one convenient place: your iPhone. Use the accompanying reports to give your doctor a good overview of your blood pressure.

» **Elevate — Brain Training (free 1-week trial, then $39.99/year subscription):** Ready to give your noggin a workout? Elevate, an App Store Editors' Choice for best app, uses a variety of more than 40 games carefully crafted to strengthen and develop your mental skills.

» **Goodreads (free):** If you're a reader, this is an app you won't want to be without. Another extremely well-rated app, Goodreads will keep you up-to-date on the latest releases, and you can browse reading lists and reviews from thousands of other users.

» **GoodRx: Prescription Saver (free):** This app is a gem when it comes to finding the lowest prescription prices in town. It's saved users untold amounts of money as opposed to simply going to your same old pharmacy and hoping you're getting the best price. GoodRx will even find coupons for you to use, and you won't have to clip them out of a flyer or newspaper advertisement!

» **Mint: Budget & Expense Manager (free, with in-app purchases available):** This highly rated financial app (almost a perfect 5 stars with nearly 750,000 reviews in the App Store!) helps you manage all your finances in one easy-to-use tool. Shown in **Figure 14-2**, Mint allows you to view account balances, track spending and bills, create budgets, and more!

» **Nike Training Club (free) and Nike Run Club (free):** Use these handy utilities to help design personalized workouts, see step-by-step instructions to learn new exercises, and watch video demonstrations. The reward system in these apps may just keep you going toward your workout and running goals.

» **Procreate Pocket ($4.99):** If you're a serious artist, you will love Procreate Pocket. It's perhaps the most complete painting, sketching, and illustration app out there.

» **Skype (free):** Make internet calls to your friends and family for free. While your iPhone comes with FaceTime, some members of your circle may not have iPhones, so in those instances Skype could be the best way to communicate.

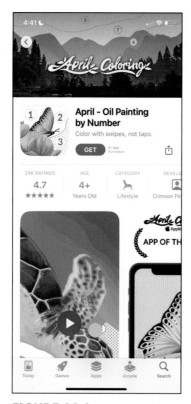

FIGURE 14-1

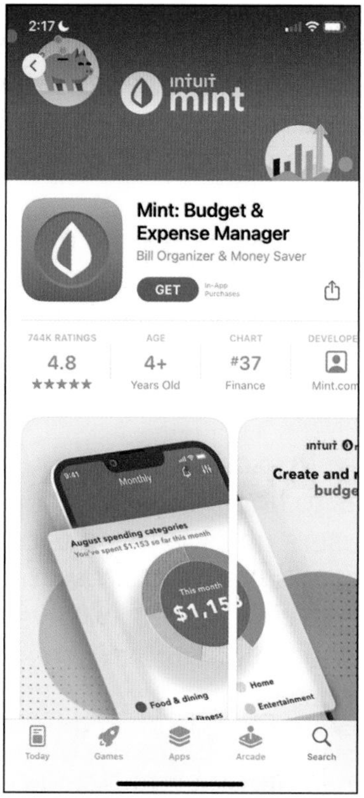

FIGURE 14-2

» **Sudoku (free):** If you like this mental logic puzzle in print, try it out on your iPhone. It has three lessons and several levels ranging from easiest to nightmare, making it a great way to make time fly almost anywhere. You can choose from several sudoku apps, so find the one that's right for you.

» **Tasty (free):** If you enjoy cooking and are always on the lookout for new recipes to try, Tasty is an excellent app to check out. Thousands of recipes are at your fingertips, and you won't need to thumb through a recipe book to find them.

» **Travelzoo Hotel & Travel Deals (free):** Get great deals on hotels, airfare, rental cars, entertainment, and more. This app also offers tips from travel experts.

>> **WordBrain — Classic Word Puzzle (free, with in-app purchases available):** Everyone loves word puzzles, and now you can have them on your iPhone instead of carrying a book and pen around wherever you go.

Note that you can work on documents using apps in the cloud. Use Keynote, Numbers, Pages, and more apps to get your work done from any device. See Chapter 4 for more about using iCloud Drive.

Search the App Store

The App Store is chock full of app goodness, but it's a lot to just browse through. Thankfully, Apple provides a handy search tool to help you find the apps you need. Here's how to use it:

1. Tap the App Store icon on the Home screen.

   By default, the first time you use App Store it will open to the Today tab, shown in **Figure 14-3**.

2. Find an app using one of the following techniques:

   - Scroll downward to view featured apps and articles, such as The Daily List and Our Favorites.

   - Tap a category to see more apps in it.

   - Tap the Apps icon at the bottom of the screen to browse by the type of app you're looking for, or search by categories, such as Lifestyle or Medical, as shown in **Figure 14-4**. To see the list of categories, tap the See All button in the Top Categories section; you may have to scroll quite a bit to get to that section.

   - Tap the Games icon at the bottom of the screen to see the newest releases and bestselling games. Explore by either Paid apps or free apps, by categories, and by special subjects, such as What We're Playing Today and New Games We Love.

   - Tap Search at the bottom right of the screen and then tap in the search field. Enter a search term, and tap the result you want to view.

FIGURE 14-3

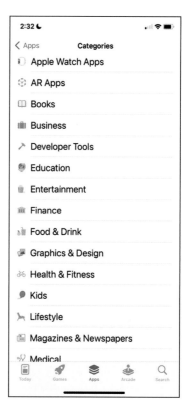

FIGURE 14-4

Get Applications from the App Store

Buying or getting free apps requires that you have an iTunes account, which I cover in Chapter 4. After you have an account, you can use the saved payment information there to buy apps or download free apps with a few simple steps.

1. With the App Store open, tap the Apps icon and then tap the See All button in the Top Free Apps section, as shown in **Figure 14-5**.

2. Tap the Get button (or the price button, if it's a paid app) for an app that appeals to you, or simply tap the app's icon if you want more information.

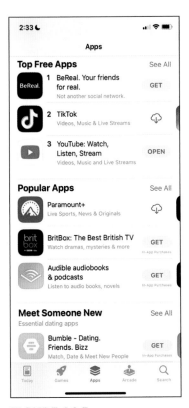

FIGURE 14-5

If you already have the app and an update is available, the Get button changes to Update. If you've already downloaded the up-to-date version of the app, the button will say Open. If you've previously downloaded the app but it's no longer on your iPhone or you've downloaded the app on another device that's signed into the same Apple ID, the icon looks like a cloud with a downward arrow; tap to download it again.

To get a paid app, you tap the same button, which is then labeled with a price.

TIP

If you've opened an iCloud account, you can set it up so that anything you purchase on your iPhone is automatically pushed to other Apple iOS devices and your iTunes library, and vice versa. See Chapter 4 for more about iCloud.

A sheet opens on the screen, listing the app and the iTunes account being used to get or purchase the app.

3. To complete the download or purchase of the app, tap Enter Password at the bottom of the sheet, tap the Password field, and then enter the password.

Alternatively, you simply only need to use Touch ID or Face ID (for iPhone models without a Home button) to approve the download or purchase. The Get (or price) button changes to the Installing button, which looks like a circle; the thick blue line on the circle represents the progress of the installation.

The app downloads and can be found on one of the Home screens. If you purchase an app that isn't free, your payment source or gift card is charged at this point for the purchase price.

Out of the box, only preinstalled apps are located on the first iPhone Home screen, with a few (such as Files and Translate) located on the second Home screen. Apps that you download are placed on available Home screens, and you have to scroll to view and use them; this procedure is covered later in the chapter. See the next task for help in finding your newly downloaded apps.

Organize Your Applications on Home Screens

By default, the first Home screen contains preinstalled apps, and the second contains a few more preinstalled apps. Other screens are created to contain any apps you download or sync to your iPhone. At the bottom of any iPhone Home screen (just above the dock), dots indicate the number of Home screens you've filled with apps; a solid dot specifies which Home screen you're on now, as shown in **Figure 14-6**.

If you see a search field instead of the aforementioned dots, simply swipe right or left on your Home screen to reveal them.

1. Press the Home button or swipe up from the bottom of the screen (for iPhone models without Home buttons) to open the last displayed Home screen.

2. To move to the next Home screen, flick your finger from right to left. To move back, flick from left to right.

3. To reorganize apps on a Home screen, press and hold down on any app on the screen. Keep your finger pressed down on the screen until the app icons begin to jiggle (see **Figure 14-7**), even if a menu pops up on screen (ignore it for our purposes here).

Dots indicating the number of Home screens

Screen you're on

FIGURE 14-6   FIGURE 14-7

If you release the screen when a menu pops up, simply tap the Edit Home Screen option to make the icons dance again.

TIP

4. To move an app to another location on the screen, hold down on the app icon and drag it.

TIP

While the apps are jiggling, you can move an app from one Home screen to another by dragging the app to the left or right. You may have to pause at the edge of the screen before it will switch over.

5. To stop all those icons from jiggling, press the Home button! For iPhone models without a Home button, tap the Done button (upper-right corner of the screen) or simply swipe up from the bottom of the screen.

TIP

You can use the multitasking feature for switching between apps easily. Press the Home button twice and you'll see a parade of open apps. For iPhone models without a Home button, swipe up from the bottom of the screen and keep your finger on the screen until App Switcher opens. Swipe right or left to scroll among the apps and tap the one you want to go to. You can also swipe an app upward from App Switcher to close it. This feature is especially helpful if an app isn't behaving as expected.

Organize Apps in Folders

iPhone lets you organize apps in folders so that you can find them more easily. The process is simple:

1. Press and hold down on an app (continue to hold down even if a menu pops up on the screen) until all apps start jiggling.

2. Drag one app on top of another app.

   The two apps appear in a box with a placeholder name in a box above them (see **Figure 14-8**).

3. To change the name, tap the placeholder name, and the keyboard appears.

4. Tap the Delete key to delete the placeholder name, and then type one of your own.

5. Tap the Done key and then tap anywhere outside the box to close it.

FIGURE 14-8

6. Press the Home button (or swipe up from the bottom of the screen for iPhone models without a Home button) to stop the icons from dancing around.

You'll see your folder on the Home screen where you began this process.

Here's a neat trick that allows you to move multiple apps together at the same time. Follow these steps (you'll probably need to place your iPhone on a flat surface to pull this off):

1. Press and hold down on the first app you'd like to move.

When the apps are jiggling, you're ready for the next step (but you'll want to move right along before the jiggling stops and you have to start over).

2. While holding down on the app, move it just a bit so that it's slightly no longer in its original place.

3. With your free hand, tap the other app(s) you'd like to move along with the first app.

   As you tap additional apps, their icons attach themselves to the first app.

4. When you've selected all your apps, drag them to their new location.

   They'll all move together in a caravan!

Delete Apps You No Longer Need

When you no longer need an app you've installed, it's time to get rid of it. Not only is it just good practice not to clutter your iPhone with unused apps, it also frees valuable storage space. You can also remove most of the preinstalled apps that are part of iOS 16.

TIP

If you use iCloud to push content across all Apple iOS devices, deleting an app on your iPhone won't affect that app on other devices.

1. Display the Home screen that contains the app you want to delete.

TIP

If you remove an app that comes with iOS 16 and decide later that you need it again, you can find the app in the App Store and reinstall it. Some apps, like Clock, will simply be removed from the Home Screen, not deleted.

2. Press and hold down on the app until all the apps begin to jiggle.

3. Tap the delete icon (— in a gray circle) in the upper-left corner of the app you want to delete. A confirmation like the one shown in **Figure 14-9** appears.

4. Tap Delete App to proceed with the deletion.

   If you want the app to remain in App Library, tap the Remove from Home Screen button instead. For more on App Library, see Chapter 2.

FIGURE 14-9

TIP

Don't worry about wiping out several apps at one time by deleting a folder. When you delete a folder, the apps that were contained in the folder are not deleted but placed back on a Home screen where space is available, and you can easily find the apps using the Search feature.

Offload Apps to Keep Data

When you delete an app from your iPhone, usually you're simultaneously deleting its data and documents. However, you can also delete an app without removing its data and documents. This feature is called offloading. If you find later that you'd like to revisit the app, simply download it again from the App Store and its data and settings will be retained.

To offload apps, follow these steps:

TECHNICAL STUFF

1. Open Settings and go to General⇨iPhone Storage.

   You may need to wait a short amount of time for content to load.

2. You can allow your iPhone to automatically offload unused apps as storage gets low, or you can offload individual apps manually:

   • To automatically offload unused apps, scroll down the screen if necessary to the Offload Unused Apps option (if you don't see it, tap Show All next to Recommendations), and then tap the Enable button, as shown in **Figure 14-10**.

     To disable the feature, go to Settings⇨App Store, scroll nearly to the bottom of the page, and then toggle the Offload Unused Apps switch off (white).

   • To offload an individual app, scroll down and tap the app, and then tap the Offload App option, shown in **Figure 14-11**; tap Offload App again to confirm. The offloaded app's icon appears dimmed on your iPhone's Home screen, indicating that the app is not loaded but its data still is.

3. You can restore an app by either tapping its icon (which will now display a cloud next to its name) or by reinstalling it from the App Store.

TIP

Often, most of your iPhone's memory is taken by the data used in apps, not by the apps themselves. Offloading apps is a great idea if the app itself is of a significant size. Otherwise, it may not be handy unless you're just super-strapped for space. It would be better to delete both the app and its data if space is at a premium on your device.

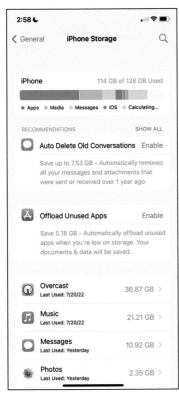

FIGURE 14-10

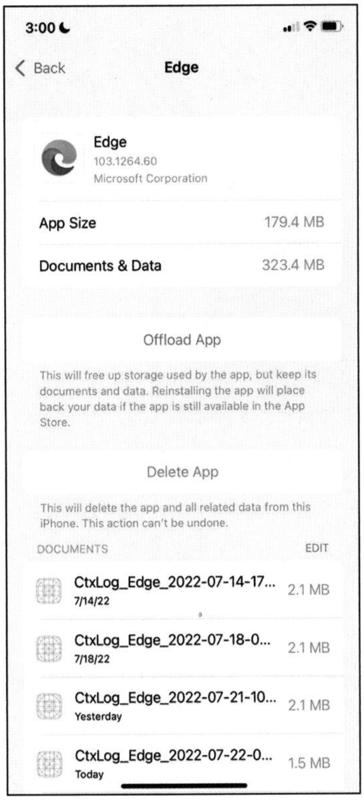

FIGURE 14-11

Update Apps

App developers update their apps all the time, so you might want to check for those updates, especially after you apply an iOS update. The App Store icon on the Home screen displays the number of available updates in a red badge. To update apps, follow these steps:

1. Tap the App Store icon on the Home screen.

2. Tap the account icon, in the upper right of the Apps screen (like the one in **Figure 14-12**).

   If updates are available, the account icon will display a red badge with a number. If you don't see a badge with a number, no updates are available.

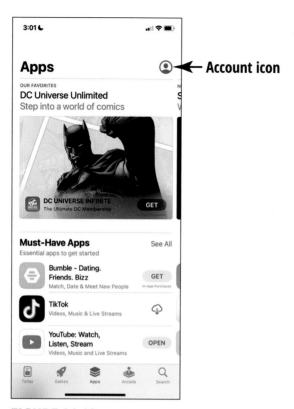

3:01

Apps

Account icon

OUR FAVORITES
DC Universe Unlimited
Step into a world of comics

DC UNIVERSE INFINITE
The Ultimate DC Membership

GET

Must-Have Apps   See All
Essential apps to get started

Bumble - Dating.
Friends. Bizz   GET
Match, Date & Meet New People   In-App Purchases

TikTok
Videos, Music & Live Streams

YouTube: Watch,
Listen, Stream   OPEN
Videos, Music and Live Streams

Today   Games   Apps   Arcade   Search

FIGURE 14-12

3. Scroll down to the Available Updates section and tap the Update button for any item you want to update (see **Figure 14-13** for examples). To update all at once, tap the blue Update All button to the left.

Note that if you have Family Sharing turned on, you'll see a Family Purchases folder that you can tap to display apps shared across your family's devices.

TIP

If you choose more than one app to update instead of downloading apps sequentially, several items will download simultaneously.

4. If you are asked to confirm that you want to update or to enter your Apple ID, do so. Then tap OK to proceed. If you are asked to confirm that you are over a certain age or to agree to terms and conditions, scroll down the terms dialog and, at the bottom, tap Agree.

You see the progress of the download.

Update everything

Update an individual app

FIGURE 14-13

TIP

If you have an iCloud account that you have activated on several devices and update an app on your iPhone, any other Apple iOS devices are also updated automatically and vice versa.

Your iPhone can update iOS automatically as updates become available. To enable this feature, go to Settings⇨General⇨Software Update⇨Automatic Updates. Toggle the Download iOS Updates, Install iOS Updates, and Security Responses & System Files switches on (green) or disable by toggling them off (white).

Sometimes Apple offers iPhone users the ability to install beta software for iOS, which is software that's not been fully tested and could cause problems with features you've become accustomed to. Don't worry because Apple won't force this on you; you have to opt-in to Apple's Beta Software Program and jump through a series of hoops to use beta software. If you decide to use the beta software, you should not use Automatic Updates until you've researched the latest version to make sure features you rely on still function correctly. Honestly, unless you have a compelling reason to enroll in the beta software program (for example, if you're an app developer or simply like living life on the razor's edge), I advise against it.

GAMING WITH APPLE ARCADE

Apple Arcade is an Apple service that allows unlimited gaming for a monthly fee of $4.99, after a free one-month trial. A subscription grants access to a ton of top-flight games that you can play online or download for offline gaming. Up to six family members can use a single subscription, so Apple Arcade is a great way for families to interact and have fun. Games can be played on iPhone, iPad, Apple TV, and Macs. You can even continue a game across devices, jumping from one device to another! The games in Apple Arcade are state-of-the-art and created by the top developers in the business. Check out www.apple.com/apple-arcade/ to sign up.

Apple Arcade is also available as part of Apple One, which is a bundle of Apple services. If you subscribe to multiple Apple services (such as Apple TV+ or Apple Fitness+), you'll save money by combining your individual subscriptions into an Apple One bundle. Visit www.apple.com/apple-one/ to find out more about the various Apple One subscriptions.

Enjoy, gamers!

Purchase and Download Games

Time to get your game on!

The iPhone is super for playing games, with its bright screen, portable size, and screen rotation capability as you play and track your motions. You can download game apps from the App Store and play them on your device.

TIP

Although a few games have versions for both Mac and iOS users, the majority are either macOS-version only (macOS is the operating system used on Apple's Mac computers) or iOS-version only — something to be aware of when you buy games.

1. Open the App Store.

2. Tap the Games icon at the bottom of the screen (see **Figure 14-14**).

3. Navigate the Games screen:

- Swipe from right to left to see featured apps in such categories as What We're Playing Today and Editors' Choice.

- Swipe down to find the Top Paid and Free games or to shop by categories. (Tap See All to view all available categories.)

4. Explore the list of games in the type you selected until you find something you like; tap the game to see its information screen.

5. To install a game, tap the button labeled with either Get or the price (such as $2.99).

6. When the dialog appears at the bottom of the screen, tap Purchase (if it's a paid game) or Install (if it's a free game, as shown in **Figure 14-15**), type your password in the Password field on the next screen, and then tap Sign In to download the game.

Alternatively, use Touch ID or Face ID (for iPhone models without a Home button) if it's enabled for iTunes and App Store purchases. The dialog box will display Pay with Touch ID, or you'll be prompted to double-click the side button to initiate Face ID authentication (for iPhones without a Home button).

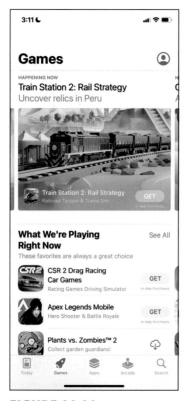

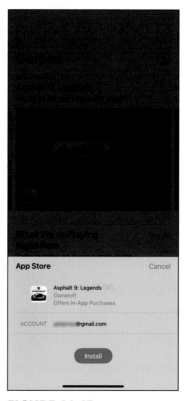

FIGURE 14-14

FIGURE 14-15

WARNING

I would warn against using Touch ID or Face ID for iTunes and App Store purchases. I know it's simpler than entering a password, but it can also make it easier for others to make purchases. In case you're wondering how that could be so, my children have tried holding my iPhone in front of my face while I was asleep in a clandestine attempt at purchasing the latest game craze with Face ID — true story. Imagine the kids in your life trying to do the same to you, and I believe you'll see where I'm coming from.

7. After the game downloads, tap the Open button to go to the downloaded game or find the game's icon on your Home screen and tap to open it.

8. Have fun!

Challenge Friends in Game Center

If you and a friend or family member have both downloaded the same games, you can challenge that person to beat your scores and even join you in a game — if the game supports Game Center interaction. You have to consult the developer's game information to find out if that's the case.

You also need to make sure your Apple ID information is registered for Game Center interaction. To do so, follow these steps:

1. Open the Settings app.

2. Scroll down a bit and tap Game Center.

3. Tap to toggle the Game Center switch on (green) and then tap Continue if prompted.

   You are logged with your Apple ID and password.

4. Toggle the Nearby Players switch on (green) if it isn't already.

   This option enables you to find and invite nearby players who connect with you over Wi-Fi or Bluetooth, assuming the game you want to play supports this functionality.

Chapter **15**

Socializing with Facebook, Twitter, and Instagram

Social media apps keep us in close digital contact with friends, family, and the rest of the world, and for some folks have become as important a digital staple as email, if not more so. Facebook, Twitter, and Instagram are some of the most popular social media apps (although there are plenty of others you should feel free to check out), and therefore I focus on obtaining and setting up these apps for use with your iPhone in this chapter. You'll find that other social media apps are as simple to set up as these.

Facebook is a platform for sharing posts about your life, with or without photos and video, and allows you to be as detailed as you please in your posts. Twitter, on the other hand, is meant to share information in quick bursts, allowing users only 280 characters in which to alert you to their latest comings and goings. Instagram is basically a photo-sharing app, allowing you to add captions to personalize your pictures.

A Few Social Media Dos and Don'ts

Social media, like most things, has its pros and cons, its upsides and downsides. While you can connect with old friends, swap stories with others, and share vacation pics of the family, you're also in danger of being preyed upon by cyber thugs and other ne'er-do-wells prowling the internet. This short list (it's by no means exhaustive) of dos and don'ts will help keep you safe on social media:

» Do connect with family and friends, but keep your social media circle close. It's great to reconnect with people you've not seen a while, but branching out too far can lead to mischief by some who don't know you.

» Do use strong and unique passwords for your accounts. This safeguard makes it tougher for folks with bad intentions to access your account and potentially post things in your name that you otherwise wouldn't condone.

» Do set up privacy controls for each social media account you use. You may want some people to see everything you post, but you may also want to keep certain things (such as birthdays and other personal information) closer to the vest.

» Don't ever share your social security number, credit card numbers, banking accounts, or any other financial information of any kind — period! Sharing financial information on social media apps is absolutely not the same thing as sharing that kind of information on secure sites, such as online banking or shopping sites.

» Don't type in ALL CAPS. It's considered the internet equivalent of yelling. You may mean nothing by it, but since there's no way for your interlocutor to hear or see the context of your text, they could misconstrue your intent. That's how all sorts of social media dust-ups get started.

» Don't accept a friend request from someone you are already friends with on a social media platform. If you're already friends with them on the platform, it's a good indication something fishy's going on. Send a private message to the friend or get in contact with them through some other way to confirm that they've sent the friend request.

» Don't believe everything you read! If something sounds too crazy to be true, it probably is. It's always best — and I do mean always — to research the topic via multiple trustworthy and varied sources before commenting on it in social media environs.

TIP

The deluge of false news on social media is beginning to take its toll on society, and social media developers are taking steps to combat the misuse of their platforms. For example, Facebook tries to flag demonstrably false articles and posts while still giving you the option to view them, as is your right.

» Don't advertise that you're on vacation. Wait until you return to post pictures of your dream trip. If someone with ill intent knows that you're away, they might take advantage of the opportunity to pay your home an unannounced and unwanted visit.

Find and Install Social Media Apps

To begin using social media apps, you first need to find and install them on your iPhone. I focus on Facebook, Twitter, and Instagram in this chapter because they're currently three of the most popular social media apps, and frankly, I don't have the space here to discuss more. To find and install the apps using the App Store, follow these steps:

1. Open the App Store.

2. Tap the Search icon at the bottom of the screen.

3. Tap the search field and enter Facebook, Twitter, Instagram, or any other social media app you might be interested in.

4. To download and install the app, tap the button labeled Get or with the price (such as $2.99) or that looks like a cloud (if you've downloaded the app before).

5. When the dialog appears at the bottom of the screen, tap Purchase (if it's a paid app) or Install (if it's free), type your password in the Password field on the next screen, and then tap Sign In to download the app.

   Alternatively, use Touch ID if you have it enabled for iTunes and App Store purchases. The dialog will display Pay with Touch ID if you do. iPhone models without a Home button can use Face ID instead of Touch ID. When you're prompted to pay (with Face ID enabled), double-click the side button and glance at your iPhone to initiate payment.

   The app will download and install on one of your Home screens.

Create a Facebook Account

You can create a Facebook account from within the Facebook app.

If you already have a Facebook account, you can simply use that account information to log in.

To create an account in the Facebook app, follow these steps:

1. Launch the newly downloaded Facebook app.

2. Tap the Create New Account button near the bottom of the screen, as shown in **Figure 15-1**.

3. Tap Get Started and walk through the steps to complete the registration of your account.

   When finished, you'll be logged into your account in the Facebook app.

You may create a Facebook account also by visiting its website at www.facebook.com.

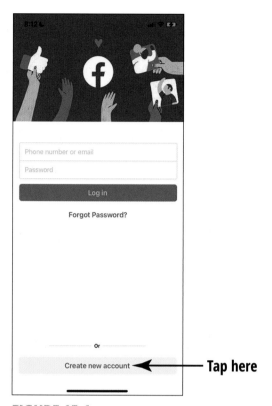

Tap here

FIGURE 15-1

Customize Facebook Settings for Your iPhone

Facebook has a few settings that you'll want to configure when entering your account information into the Settings app.

1. Open the Settings app.

2. Tap Facebook.

3. Toggle the switches shown in **Figure 15-2** on (green) or off for the following options:

   - *Background App Refresh:* This option allows Facebook to refresh its content in the background (when you aren't using the app).

- *Cellular Data:* Turning this on allows Facebook to refresh itself and enables you to post updates when you aren't connected to a Wi-Fi network.

FIGURE 15-2

4. Tap the remaining items in the Allow Facebook to Access section (Photos, Bluetooth, Local Network, Siri & Search, and Notifications) to customize how Facebook can interact with these iOS 16 features.

   For example, tap Siri & Search and allow or deny Facebook to use Ask Siri.

WARNING

If you are a frequent Facebook user and tend to upload a lot of photos and videos, consider toggling the Cellular Data switch off. If you have an account with a cellular provider that provides a limited amount of data, you could be in danger of exceeding your data allotment (which translates into a more expensive bill) if you're a heavy Facebook user.

Create a Twitter Account

To create an account in the Twitter app, follow these steps:

1. Open the Twitter app by tapping its icon.

2. Tap the blue Create Account button in the middle of the screen, as shown in **Figure 15-3**. Or, if you want to create a Twitter account that's linked to your Google account or Apple ID, click either Continue with Google or Continue with Apple, respectively.

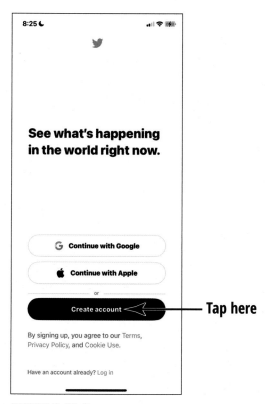

FIGURE 15-3

TIP

If you already have a Twitter account, tap the tiny Log In link at the bottom of the screen to log in.

3. Answer the questions the app asks to help you create your account.

When you're done, the app will log you into your new account.

TIP

You can create an account on the Twitter website at www.twitter.com. You can also configure options for Twitter in the Settings app, as detailed for Facebook in the preceding section.

Create an Instagram Account

To create an account in the Instagram app:

1. Open the Instagram app by tapping its icon.

2. Tap the blue Create New Account button in the middle of the screen.

TIP

If you already have an Instagram account, tap the small Log In button to access it.

3. Answer the question the app asks to create your account.

When completed, the app will log you into your new account.

TIP

You can create an account also on the Instagram website at www.instagram.com. You may also configure options for Instagram in the Settings app, as explained for Facebook earlier in this chapter. (Note that the Instagram options are more limited than those for Facebook.)

Enjoying Media

IN THIS PART . . .

Shopping for music, movies, and more

Reading electronic books

Listening to audio

Taking and sharing photos and videos

Getting directions

Chapter **16**

Shopping the iTunes Store

The iTunes Store app lets you easily shop for music, movies, and TV shows. As Chapter 17 explains, you can also get electronic and audio books via Apple's Books app.

In this chapter, you learn how to find content in the iTunes Store. You can download the content directly to your iPhone or you can download the content to another device and then sync it to your iPhone. With the Family Sharing feature, which I cover in this chapter, as many as six people in a family can share purchases using the same credit card. Finally, I cover a few options for buying content from other online stores and using Apple Pay to make real-world purchases using a stored payment method.

TIP

I cover opening an iTunes account and downloading iTunes software to your computer in Chapter 4.

Explore the iTunes Store

Visiting the iTunes Store from your iPhone is easy with the iTunes Store app.

TIP

If you're in search of other kinds of content, the Podcasts app (discussed in detail in Chapter 18) allows you to find and then download podcasts to your phone.

To check out the iTunes Store, follow these steps:

1. Go to your Home screen and tap the iTunes Store icon.

2. Tap the Music icon (if it isn't already selected) in the row of icons at the bottom of the screen. Swipe up and down the screen and you'll find several categories of selections, such as New Music, Today's Hits, and Recent Releases.

   These category names change from time to time.

3. Flick your finger up to scroll through the featured selections or tap the See All button to see more selections in any category, as shown in **Figure 16-1**.

TIP

   The navigation techniques in these steps work essentially the same in any of the content categories (the icons at the bottom of the screen), which are Music, Movies, and TV Shows.

4. Tap the Charts tab at the top of the screen.

   The screen displays lists of bestselling songs, albums, and music videos in the iTunes Store.

5. Tap any listed item to see more detail, as shown in **Figure 16-2**. Hear a brief preview by tapping the number to the left of a song.

TIP

If you want to use the Genius playlist feature, which recommends additional purchases based on the contents of your library in the iTunes app on your iPhone, tap the More icon at the bottom of the screen, and then tap Genius. If you've made enough purchases in iTunes, song and album recommendations appear based on those purchases as well as the content in your iTunes Match library (a fee-based service discussed in Chapter 18), if you have one.

Tap See All

FIGURE 16-1   FIGURE 16-2

Buy a Selection

Once you have found that new tune you can't live without, or have rediscovered those older tracks that bring back so many memories, you may decide to purchase them for your collection.

1. When you find an item that you want to buy, tap the button that shows the price (if it's a selection available for purchase, as shown in **Figure 16-3**) or the button labeled Get (if it's a selection available for free).

TIP

If you want to buy music, you can open the description page for an album and tap the album price, or buy individual songs rather than the entire album. Tap the price for a song and then proceed to purchase it.

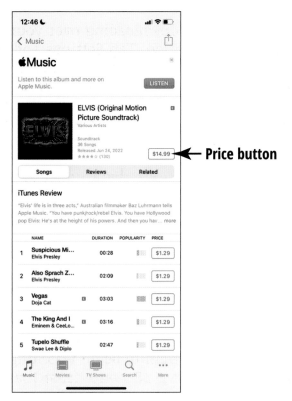

Price button

FIGURE 16-3

2. When the dialog appears at the bottom of the screen, tap Purchase, type your password in the Password field on the next screen, and then tap Sign In to buy the item.

Alternatively, use Touch ID if you have it enabled for iTunes and App Store purchases. The dialog displays Pay with Touch ID if you do. Also, users of an iPhone without a Home button can use Face ID; when asked to pay, double-click the side button and look at your iPhone to authenticate.

The item begins downloading and the cost, if any, is automatically charged against your account.

3. When the download finishes, tap OK in the Purchase Complete message.

You can now view the content using the Music or TV app, depending on the type of content.

If you aren't near a Wi-Fi hotspot, downloading over your cellular network might be your only option. Tap Settings and then tap Cellular. Scroll down to the Cellular Data section, find iTunes Store, and set the switch on (green). Go through the list of apps while you're there and enable or disable apps as you see the need, especially if you allow others to use your iPhone and are concerned that they may consume large quantities of cellular data.

Music files are large, usually several megabytes per track. You could incur hefty data charges with your cellular provider if you run over your allotted data.

Rent Movies

In the case of movies, you can either rent or buy content. If you rent, which is often less expensive but only a one-time deal, you have 30 days from the time you rent the item to begin watching it. After you have begun to watch it, you have 48 hours from that time to watch it on the same device, as many times as you like.

Music files are large, but movie files are much, much larger. I highly recommend downloading movie files only via Wi-Fi, if possible. Plus, they'll typically download much more quickly.

Here's how to rent a movie:

1. With the iTunes Store open, tap the Movies icon.

2. Locate the movie you want to rent and tap it, as shown in **Figure 16-4**.

3. In the detailed description of the movie that appears, tap the Rent button (if it's available for rental), as shown in **Figure 16-5**.

4. When the dialog appears at the bottom of the screen, tap Rent, type your password in the Password field on the next screen, and then tap Sign In to rent the item.

Buy button   **Rent button**

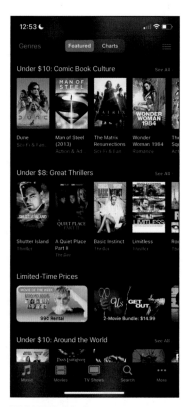

FIGURE 16-4

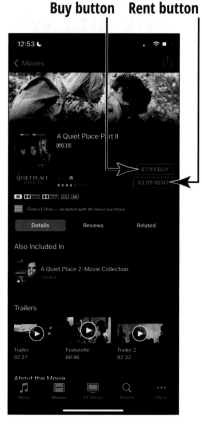

FIGURE 16-5

Alternatively, use Touch ID if you have it enabled for iTunes and App Store purchases. The dialog displays Pay with Touch ID if you do. If your iPhone doesn't have a Home button, you can use Face ID; when asked to pay, double-click the side button and look at your iPhone to authenticate.

The movie begins to download to your iPhone immediately, and your account is charged the rental fee. After the download is complete, you can use the TV app to watch it. (See Chapter 20 to read about how this app works.)

You can also download content to your computer and sync it to your iPhone. Refer to Chapter 4 for more about this process.

TIP

Use Apple Pay and Wallet

Apple Pay, fancied as a mobile wallet, uses Touch ID (or Face ID if your iPhone doesn't have a Home button) to identify you and credit cards you've stored at the iTunes Store to make payments via a feature called Wallet.

Your credit card information isn't stored on your phone, and Apple doesn't know a thing about your purchases. In addition, Apple considers Apple Pay safer than paying with a credit card at the store because the store cashier doesn't even have to know your name.

You can store more than just credit cards in Apple Wallet. Apple Wallet supports debit cards, ID cards, electronic tickets (for movies, as one example), transit fare cards, and more.

To set up Apple Pay, go to Settings and tap Wallet & Apple Pay. Add information about a credit card and then double-tap the Home button (side button for iPhone models without a Home button) when the lock screen is displayed to initiate a purchase. You can also make settings from the Wallet app itself. For more information on Apple Pay, check out `www.apple.com/apple-pay`, and to find more on Apple Wallet, visit `www.apple.com/wallet`.

Set Up Family Sharing

Family Sharing allows as many as six people in your family to share whatever anyone in the group has purchased from the iTunes, Apple Books, and App Stores even though you don't share Apple IDs. Your family must all use the same credit card to purchase items (tied to whichever Apple ID is managing the family); you can approve purchases by children under 13 years of age. (This age can vary depending on your country or region.) You can also share calendars, photos, and a family calendar.

Start by turning on Family Sharing.

1. Tap Settings and then tap Apple ID at the top of the screen.

2. Tap Set Up Your Family.

3. Tap Get Started. On the next screen, you can add a photo for your family. Tap Continue.

4. On the Share Purchases screen, tap Share Purchases from a different account to use another Apple account.

5. Tap Continue and select the payment method you want to use. Tap Continue.

6. On the next screen, tap Add Family Member. Enter the person's name (assuming that this person is listed in your contacts) or email address.

   An invitation is sent to the person's email. When the invitation is accepted, the person is added to your family.

7. To add additional family members later, go to Settings ⇨ *Apple ID account name* ⇨ Family Sharing, and tap the blue Add Member button in the upper-right corner. Follow the steps as prompted.

TECHNICAL STUFF

The payment method for this family is displayed under Purchase Sharing in this screen. All those involved in a family have to use a single payment method for family purchases.

There's also a link called Create a Child Account. When you tap this link and enter information to create the ID, the child's account is automatically added to your Family and retains the child status until the child turns 13. If a child accesses iTunes to buy something, a prompt appears to ask permission. You get an Ask to Buy notification on your phone in the Messages app and via email. You can then accept or decline the purchase, giving you control over your child's spending in the iTunes Store.

Chapter **17**

Reading Books

A traditional e-reader is a device that's used primarily to read the electronic versions of books, magazines, and newspapers. Apple's free app that turns your iPhone into an e-reader is Apple Books. This app also enables you to buy and download e-books and audiobooks from the Apple Book Store (offering millions of books and growing by the day).

In this chapter, I help you discover the options available for reading material and how to buy books. You also learn how to navigate a book or periodical and adjust the brightness and type. If you've used the Books app in prior iOS versions, you'll find that book navigation and customization settings have changed a bit. But no worries — I cover them all in the upcoming pages.

Find Books with Apple Books

The Apple Books app isn't much to look at unless it's populated with books for you to peruse and enjoy. As always, Apple has already thought this through and provided its Book Store for you to explore.

To shop using Apple Books:

1. Tap the Books icon to open the app.

2. Tap the Book Store icon at the bottom of the screen.

3. In the Book Store, shown in **Figure 17-1**, featured titles and suggestions (based on your past reading habits and searches) are shown. You can do any of the following to find a book:

    - Tap the Search icon in the bottom right of the screen, tap in the search field that appears, and then type a search word or phrase using the onscreen keyboard.

    - Swipe left or right to see and read articles and suggestions for the latest books in various categories, such as New (shown in **Figure 17-2**) and Bestsellers (current popular titles).

FIGURE 17-1

FIGURE 17-2

- Scroll down the Book Store's main page to see links to popular categories of books, as shown in **Figure 17-3**. Tap a category to view those selections.

- Scroll down to Top Charts to view both paid and free books listed on top bestseller lists. Tap the See More Paid or See More Free button in Top Charts to focus on books that are the latest hits in those charts.

- Swipe further down the screen to find a list of genres. Tap All Genres to see everything the Book Store has to offer.

- Back on the main screen of the Book Store, tap Browse Sections under Book Store (at the top of the Book Store screen, if you've scrolled down) to open the Browse Sections menu, shown in **Figure 17-4**. From here you can easily scroll up and down the screen to browse book store sections and genres.

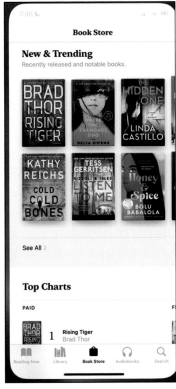

FIGURE 17-3

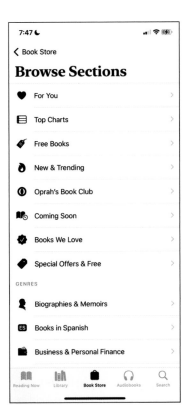

FIGURE 17-4

- Tap a suggested selection or featured book to read more information about it.

TIP

Many books let you download free samples before you buy. You get to read several pages of the book to see whether it appeals to you, and it doesn't cost you a dime! Look for the Sample button when you view a book's details. (The button usually appears below the price of the book.)

Buy Books

If you've set up an account with iTunes, you can buy books at the Apple Book Store using the Apple Books app. (See Chapter 4 for more about iTunes.)

1. Open Apple Books, tap Book Store, and begin looking for a book.

2. When you find a book in the Book Store, you can buy it by tapping it and then tapping the Buy|price button or the Get button (if it's free), as shown in **Figure 17-5**.

   If you'd like to keep this book in mind for a future purchase, tap the Want to Read button. Or to download a few pages to read before you commit your hard-earned dollars, tap Sample.

3. When the dialog appears at the bottom of the screen, tap Purchase, type your password in the Password field on the next screen, and then tap Sign In to buy the book.

   Alternatively, use Touch ID (or Face ID for iPhone models without Home buttons) if you have it enabled for iTunes and App Store purchases. The dialog displays Pay with Touch ID. iPhone users without a Home button see a prompt on the right side of the screen to double-click the side button and glance at the iPhone to authenticate with Face ID.

   The book begins downloading, and the cost, if any, is automatically charged to your account.

4. When the download finishes, tap OK in the Purchase Complete message. Find your new purchase by tapping the Library icon at the bottom of the screen.

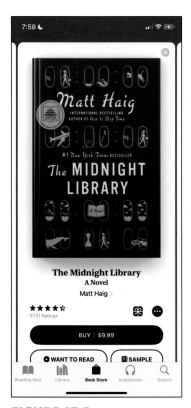

FIGURE 17-5

TIP

Books that you've downloaded to your computer can be accessed from any Apple device through iCloud. Content can also be synced with your iPhone by using the Lightning-to-USB or Lightning-to-USB-C cable (depending on which type your iPhone uses) and your iTunes account, or by using the wireless iTunes Wi-Fi Sync setting on the General Settings menu. See Chapter 4 for more about syncing.

Navigate a Book

Getting around in Apple Books is half the fun!

1. Open Apple Books and, if your library isn't already displayed, tap the Library icon (four books on a shelf) at the bottom of the screen.

2. Tap a book to open it. The book opens to its title page or the last spot you read on any compatible device.

3. Take any of these actions to navigate the book:

- *To go to the book's table of contents:* Tap the Settings icon at the bottom right of the page (see **Figure 17-6**), tap the Contents button/slider (shown in **Figure 17-7**), and then tap the name of a chapter to go to it.

- *To turn to the next page:* Place your finger anywhere along the right edge of the page and tap or flick to the left.

- *To turn to the preceding page:* Place your finger anywhere on the left edge of a page and tap or flick to the right.

- *To move to another page in the book:* Touch and hold down on the Contents button/slider (refer to Figure 17-7), drag to the right or left until you find the page number you want to move to, and then remove your finger from the Contents button/slider.

- *To bookmark a page:* Tap the Settings icon at the bottom right of the page (refer to Figure 17-6), and then tap the bookmark icon (ribbon) in the lower right (refer to Figure 17-7). The bookmark icon turns red when the page is bookmarked. You can view a list of bookmarks by tapping the Settings icon and then tapping the Bookmarks & Highlights button. Then just tap a bookmark from the list to be whisked away to the bookmarked page.

Tap the Search Book button to search the current book for keywords or by page number, and then tap the result you want when listed.

Return to the library to view another book at any time by tapping the X icon in the upper-right corner of the screen. You can also simply swipe down from the top of the screen to close the current book.

You can lock the orientation of your book's text in portrait or landscape mode (depending on how you're holding your iPhone) by tapping the lock rotation icon in Settings (refer to Figure 17-7).

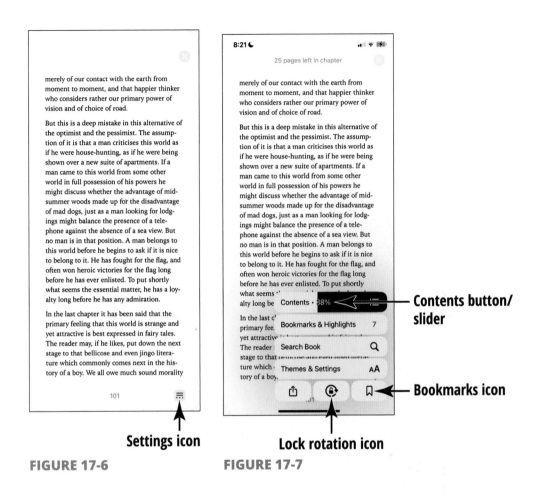

FIGURE 17-6

FIGURE 17-7

Contents button/slider

Bookmarks icon

Settings icon

Lock rotation icon

Select and Customize Themes

You've always been able to customize the look and feel of your books in the Books app, but iOS 16 has turned up the ability to customize a notch or two. Let's check out how you can make your reading experience suit your needs and taste in the revised Books app.

The Books app comes with six predefined themes: Original, Quiet, Paper, Bold, Calm, and Focus. Each provides a combination of fonts and background colors curated for a particular reading experience (as their names skillfully suggest). Feel free to select one of these predefined themes, or customize them to your liking.

1. With a book open, tap the Settings icon in the lower right of the screen (refer to Figure 17-6).

2. Tap the Themes & Settings button (refer to Figure 17-7).

3. Tap to select one of the predefined themes, shown in **Figure 17-8**.

   Original is the default theme. After you select a different one, the new theme is applied to all your books until you change it.

TIP

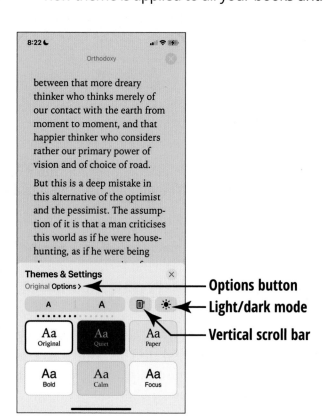

FIGURE 17-8

4. Further customize your book's appearance, if you like (refer to Figure 17-8 to locate the following items):

   - Tap the small A to decrease the font size, or tap the larger A to increase it.

   - Tap the vertical scroll bar icon to enable or disable this feature, which allows you to vertically scroll through the text of your book as opposed to swiping right or left to "turn" pages. To view the vertical scroll bar when enabled, tap the Settings icon in the lower right of your screen; the scroll bar appears to the right of the other buttons.

   - Tap the light/dark mode icon to switch between the two modes, or to allow the Books app to match your surroundings or the default settings for your iPhone.

Modify Your Book's Font

If the type on your screen is difficult for you to make out, you might want to choose a different font and perhaps use bold text for readability:

1. With a book open, tap the Settings icon in the lower right, and then tap the Themes & Settings button.

2. In the Themes & Settings sheet, tap the Options button (refer to Figure 17-8).

3. In the Options sheet, shown in **Figure 17-9**, do one or both of the following:

   - Tap the Font button, and then tap a font name to select it. The font changes on the book page.

   - Tap the Bold Text switch to enable or disable bold text.

4. Tap the Done button in the upper-right corner to save your changes and return to the Themes & Settings sheet, or tap the Cancel button in the upper-left corner to discard your changes.

FIGURE 17-9

Adjust Accessibility & Layout Options

A few new features in Books allow you to further enhance your reading experience. They can be found in the Accessibility & Layout area of the Options sheet:

1. With a book open, tap the Settings icon in the lower right, and then tap the Themes & Settings button.

2. In the Themes & Settings sheet, tap the Options button (refer to Figure 17-8).

3. In the Options sheet, tap the Customize switch in the Accessibility & Layout Options area to enable these features (refer to Figure 17-9).

4. In the Accessibility & Layout Options section, shown in **Figure 17-10**, do one, some, or all of the following:

   - Adjust the line spacing, character spacing, or word spacing by dragging their respective sliders right or left.

   - Tap to enable or disable the Full Justification switch, which, if enabled, automatically adjusts the spacing of characters and words so that the text fills the width of the screen.

   - Tap the Allow Multiple Columns switch to enable or disable the use of columns (if they're present in any of your books, that is).

5. Tap the Done button in the upper-right corner to save your changes and return to the Themes & Settings sheet, or tap the Cancel button in the upper-left corner to leave everything the way you found it.

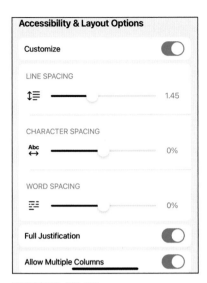

FIGURE 17-10

TIP

Don't like any of the changes you've made and want to start from scratch? Easily done, dear reader! Tap the Settings icon in the lower-right side of your book, tap the Themes & Settings button, tap the Options button, swipe to the bottom of the options (if you need to) and tap the Reset Theme button (refer to Figure 17-9 for an example), and then tap the Reset button when prompted for verification. Done!

IN THIS CHAPTER

» View the library and create playlists

» Search for music

» Play and shuffle music

» Listen with earbuds

» Check out spatial audio

» Use AirPlay

» Play music with the Radio app

» Find, subscribe to, and play podcasts

Chapter **18**

Enjoying Music and Podcasts

OS 16 includes an app called Music that allows you to take advantage of your iPhone's amazing little sound system to play your favorite music or other audio files.

In this chapter, you get acquainted with the Music app and its features that allow you to sort and find music and control playback. You also get an overview of AirPlay for accessing and playing your music over a home network or over any connected device (this also works with videos and photos). Finally, I introduce you to podcasts to enhance your listening pleasure.

View the Music Library

The library in Music contains the music or other audio files that you've placed on your iPhone, either by purchasing them through the iTunes Store or copying them from your computer. Let's see how to work with those files on your iPhone.

1. Tap the Music app, which is located in the dock on the Home screen, and then tap the Library icon at the bottom of the screen.

   The Music library appears, as shown **Figure 18-1**. You can view music by categories — Playlists, Artists, Albums, Songs, or Downloaded music — and by music you've recently added.

2. Swipe up the screen to scroll down through the Recently Added section of the Music app's Library.

3. Tap a category to see music in that category, such as the Albums category in **Figure 18-2**. Tap Library in the upper-left corner to return to the main Library screen.

TIP

The iTunes Store has several free items that you can download and use to play around with the features in Music. You can also sync content (such as Smart Playlists stored on your computer or other Apple devices) to your iPhone, and play it using the Music app. (See Chapter 4 for more about syncing and Chapter 16 for more about getting content from iTunes.)

4. To change the list of displayed categories:

   (a) Tap Edit in the upper-right corner.

   (b) Select the check box to the left of categories you'd like to use to browse your Music library, and deselect the categories you don't want to use, as shown in **Figure 18-3**. You can also drag the categories into your preferred order.

   (c) Tap Done in the upper-right corner when you're finished. Changes are reflected in the Library screen.

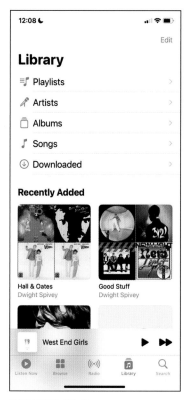

FIGURE 18-1

FIGURE 18-2

TIP

Apple offers a service called iTunes Match (visit `https://support.apple.com/en-us/HT204146` for more information). You pay $24.99 per year for the capability to match the music you've bought from other providers (and stored in the music library on your computer) to what's in Apple's Music Library. If there's a match (and there usually is), that content is added to your music library on iCloud. Then, using iCloud, you can sync the content among all your Apple devices. Is this service worth $24.99 a year? That's up to you, my friend. However, for a few bucks more, you can have the benefits of iTunes Match plus access to millions of songs across all your Apple devices by using another Apple service: Apple Music (to which I'm admittedly partial). There's more about Apple Music later in this chapter, but for more info about subscribing and what a full-blown Apple Music subscription offers, check out `www.apple.com/apple-music`.

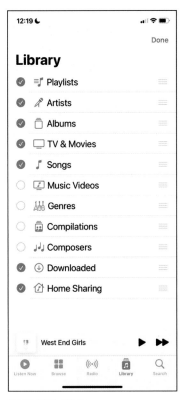

FIGURE 18-3

Create Playlists

You can create your own playlists to put tracks from various sources into collections of your choosing.

1. Tap the Playlists category at the top of the Library screen, and then tap New Playlist.

2. In the screen that appears (see **Figure 18-4**), tap Playlist Name and enter a title for the playlist. You may also tap the Description area to describe the contents of the playlist (for example, "My favorite songs from high school" or "Romantic songs for romantic times").

3. To find the tracks you're looking for, tap Add Music (you may need to swipe up to see it); search for music by artist, title, or lyrics; or browse for songs by tapping Listen Now, Browse, or Library.

4. In the list of selections that appears (see **Figure 18-5**), tap the plus sign to the right of each item you want to include (for individual songs, entire albums, or artists). Continue until you've selected all the songs you want to add to the playlist.

5. To finish adding tracks to the playlist, tap the Done button in the upper-right corner. Then tap Done (same location) on the next screen to return to the Playlists screen.

Your playlist appears in the list, and you can now play it by tapping the playlist name and then tapping a track to play it.

Tap to add album to playlist

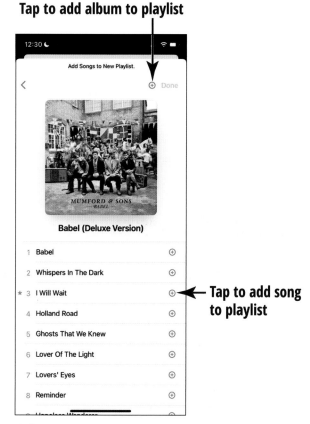

Tap to add song to playlist

FIGURE 18-4

FIGURE 18-5

Search for Music

You can search for an item in your Music library by using the Search feature.

1. With Music open, tap the Search button at the bottom of the screen.

   The Search screen appears displaying music categories, along with a search field at the top of the screen.

2. Tap the search field to display the Apple Music and Your Library tabs, as shown in **Figure 18-6**. Then tap the Your Library tab to search for songs stored on your iPhone, or tap Apple Music to search the Apple Music library.

   You can search for items in Apple Music, but you must be subscribed to the service to play selections from it.

3. Enter a search term in the search field.

   As you type, you see results, which narrow as you continue to type, as shown in **Figure 18-7**.

4. Tap an item in the results to view or play it.

TIP

In the search field, you can enter an artist's name, a lyricist's or a composer's name, a word from the item's title, or even lyrics.

TIP

To search for and play a song in your music libraries from the Home screen (without even opening the Music app), use the Search feature. From the Home screen, swipe down on the screen anywhere outside the dock and enter the name of the song in the search field (which is towards the middle of the screen). A list of search results appears. Just tap the song you're searching for (you may have to scroll down the screen a bit to see it) and you'll be whisked away to it in the Music app.

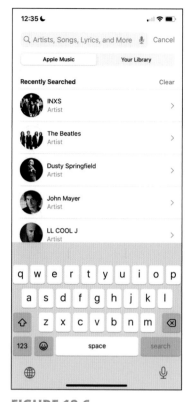

FIGURE 18-6

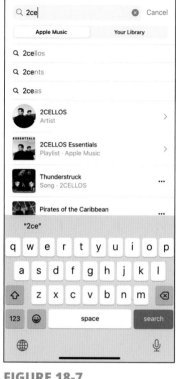

FIGURE 18-7

Play Music

Now that you know how to find your music (which will be more and more handy as your library grows), let's have some real fun by playing it!

TIP

You can use Siri to play music hands free. Just press and hold down on the Home button (or the side button if your iPhone doesn't have a Home button). When Siri appears, say something like "play 'Gypsy'" or "play 'Shockadelica' by Prince."

To play music on your iPhone, follow these steps:

1. Locate the music that you want by using the methods described in previous tasks in this chapter.

2. Tap the item you want to play.

If you're displaying the Songs category, you don't have to tap an album to open a song; you need only tap the song to play it. If you're using any other categories, you have to tap items such as albums (or multiple songs from one artist) to find the song you want to hear.

TIP

Home Sharing is a feature of iTunes and the Music app that enables you to share music among up to five devices that have Home Sharing turned on. After Home Sharing is set up via iTunes or the Sharing pane in System Preferences or System Settings (Mac users only), any of your devices can stream music and videos to other devices, and you can even click and drag content between devices using iTunes or the Music app. For more about Home Sharing, visit `https://support.apple.com/en-us/HT202190`.

3. Tap the item you want to play from the list that appears, and it will begin to play, as shown in **Figure 18-8**.

4. Tap the currently playing song title near the bottom of the screen to display playback controls, as shown in **Figure 18-9**.

5. Give the playback controls a trial run so you can see exactly how to use them when the time comes:

   - *Previous icon:* The previous icon takes you back to the beginning of the item that's playing if you tap it, takes you back to the start of the previous item if you tap it again, or rewinds the song if you press and hold down on it.

   - *Next icon:* The next icon takes you to the next item if you tap it, or fast-forwards the song if you press and hold down on it.

   - *Volume slider:* Drag the volume slider on the bottom of the screen (or press the volume buttons on the side of your iPhone) to increase or decrease the volume.

TIP

Rewinding songs with the previous icon or fast-forwarding songs with the next icon requires a bit of patience, at least for me. They both tend to jump around a bit in the process of rewinding or fast-forwarding. I much prefer dragging the playhead left or right.

   - *Pause icon:* Tap the pause icon to pause playback. Tap the play icon, which replaces the pause icon, to resume playing.

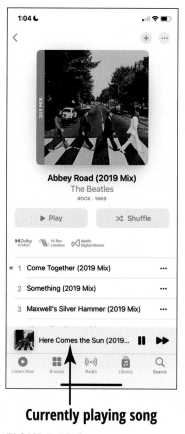

Playhead

Previous

Next

Pause

Currently playing song　　　　**Lyrics**　　**Volume**

FIGURE 18-8　　　　　　　　**FIGURE 18-9**

TIP

- *Playhead:* Tap and drag the playhead, which indicates the current playback location. Drag the playhead to the left or right to "scrub" to another location in the song.

  You can use music controls also for music playing from the lock screen.

6. Do you like to sing along but sometimes flub the words? Tap the lyrics icon in the lower left, and if the song is from the Apple Music library, the lyrics will scroll up the screen in sync with the song, as shown in **Figure 18-10**. You can swipe through the lyrics, or if you tap a lyric Music will jump to that point in the song.

TIP

If you have an Apple Music subscription, you can see lyrics for any song (assuming it does indeed have lyrics). But what if you don't have an Apple Music subscription? You can add lyrics to songs on your own if you have a Mac or PC! Using the Music app on your Mac or iTunes on a Windows-based PC, click the more icon (three dots) next to the name of a song and select Get Info from the list. Click the Lyrics tab, select the Custom Lyrics check box, add your lyrics, and then click OK. Next, copy the song to your iPhone, and the lyrics will appear when the song is played and you tap the lyrics icon.

7. If you don't like what's playing, make another selection by dragging down from the top of the playback controls screen to view other selections in the album that's playing and make a new choice.

TIP

Family Sharing allows up to six members of your family to share purchased content even if they don't share the same iTunes account. You can set up Family Sharing under iCloud in Settings. See Chapter 16 for more about Family Sharing.

FIGURE 18-10

Shuffle Music

If you want to play a random selection of the music in an album on your iPhone, you can use the shuffle feature.

1. Tap the name of the currently playing song at the bottom of the screen.

2. Tap the menu icon (three dots and three lines), in the lower right.

3. Tap the shuffle icon, located to the right of Playing Next, as shown in **Figure 18-11**.

   Your content plays in random order.

4. Tap the repeat icon (refer to Figure 18-11) to play the songs over again continuously.

FIGURE 18-11

Listen with Your Earbuds

If you've set the volume slider as high as it goes but you're still having trouble hearing the music, consider using earbuds. (If you have a model older than the iPhone 12, it came with earbuds in the box.) These cut out extraneous noises and improve the sound quality of what you're listening to.

For iPhone models older than the iPhone 7 and 7 Plus, use 3.5 mm stereo earbuds; insert them in the headphone jack at the bottom of your iPhone. If you have an iPhone 7, 7 Plus, or a newer model, you'll need earbuds that use a Lightning connector, or you can use a Lightning-to-3.5mm adaptor with standard 3.5mm headphones.

You might also look into purchasing Bluetooth earbuds, which allow you to listen wirelessly. For a top-of-the-line wireless experience, try out Apple's AirPods, AirPods Pro, or AirPods Max (go to www. apple.com/airpods for more info); they're a little pricey but getting rave reviews, and for very good reason.

Bluetooth is a tried-and-true technology, but even so, its connections and range can be flaky. To minimize the chances of a sub-par (and possibly infuriatingly frustrating) listening experience, remember that "you get what you pay for" is often accurate. As mentioned, Apple's AirPods line can be pricey. An alternative Apple product, Beats by Dr. Dre (www.beatsbydre.com), offers a wider range of earbuds and headphones, including some that aren't as expensive.

Listen with Spatial Audio

Spatial audio is a new technology (part of Dolby Atmos) that helps listeners feel as if they're sitting in the middle of the band, orchestra, or what-have-you. The difference between spatial audio and standard stereo can be eye-popping when you first hear it.

TIP For more on Dolby Atmos, such as how to tell if audio files are in the format or whether your device can play the format, visit https://support.apple.com/en-us/HT212182.

To enable Dolby Atmos for audio files that support the format:

1. Open the Settings app on your iPhone.

2. Find and tap Music.

3. Tap Dolby Atmos, and then tap Automatic or Always On to have audio files that support Dolby Atmos play in that format.

To enable spatial audio while using AirPods Pro or AirPods Max:

1. Swipe down from the top-right corner of your iPhone's screen (or up from the bottom of the screen, if your iPhone has a Home button) to open Control Center.

2. Tap and hold down on the volume slider until the screen in **Figure 18-12** opens.

Spatial Audio button

FIGURE 18-12

If the audio file that's currently playing supports Dolby Atmos and spatial audio, the Spatial Audio button will appear in the lower right (refer to Figure 18-12).

3. To enable or disable spatial audio, tap the Spatial Audio button. Then you can tap the Off button to disable spatial audio, or tap Fixed or Head Tracked to enable it in one of these two modes.

If the button is blue, spatial audio is enabled; if it's gray, spatial audio is not enabled.

Use AirPlay

AirPlay streaming technology is built into the iPhone, iPod touch, Macs, PCs running iTunes, and iPad. Streaming technology allows you to send media files from one device that supports AirPlay to be played on another. For example, you can send a movie you've purchased on your iPhone or a slideshow of your photos to be played on your Apple TV, and then control the TV playback from your iPhone. You can also send music to be played over compatible speakers, such as Apple's HomePod. (Go to www.apple.com/homepod for more information.) Check out Apple's Remote app in the App Store, which you can use to control your Apple TV from your iPhone.

To stream music via AirPlay on your iPhone with another AirPlay-enabled device on your network or in close proximity, tap the AirPlay icon (a triangle with sound waves) at the bottom center of the playback control screen while listening to a song. Then select the AirPlay device to stream the content to (see **Figure 18-13**), or choose your iPhone to move the playback back to it.

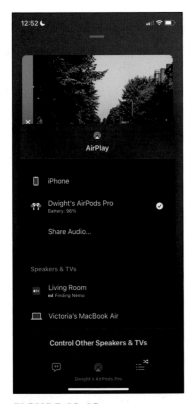

FIGURE 18-13

Play Music with Radio

You can access Radio by tapping the Radio icon at the bottom of the Music screen, as shown in **Figure 18-14**.

Swipe from left to right and you'll find lots of music specials and offers to listen to. Or swipe down to see Radio's offerings by categories such as Local Broadcasters, International Broadcasters, and Genre (see **Figure 18-15**).

TECHNICAL STUFF

At one time, the Radio feature was free in the Music app, but now it's tied extensively into Apple's Music service, which is subscription based. The Radio app offers much more when you have an Apple Music subscription, but it can still be used without one for broadcast radio stations and Apple's flagship radio station, Beats 1. However, if you want to listen to most digital stations in Radio, you'll have to become a member of Apple Music.

FIGURE 18-14

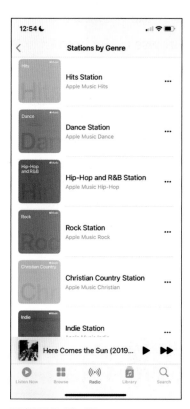

FIGURE 18-15

Find and Subscribe to Podcasts

A podcast is sort of like a radio show with the glorious difference that you can listen to it at any time. The Podcasts app is the vehicle by which you'll find and listen to podcasts on your iPhone. To search Apple's massive library of podcasts and subscribe to them (which is free, by the way), just follow these steps:

1. Tap the Podcasts icon to open the app.

2. Discover podcasts in one of the following ways:

   - Tap Browse at the bottom of the screen. You'll see podcasts featured by the good folks at Apple, as shown in **Figure 18-16**.

   - Tap Browse at the bottom of the screen, and then swipe all the way down to the Podcasts Quick Links section and tap Charts. Tap See All for Top Shows or Top Episodes and you'll find lists of the most popular podcasts. Tap All Categories in the upper-right corner to sift through the podcasts based on category (such as News, Comedy, Religion & Spirituality, Science, or Technology).

   - Tap Search and then tap the search field at the top of the screen. When the keyboard appears, type the name or subject of a podcast to see a list of results.

FIGURE 18-16

3. When you find a podcast that intrigues you, tap its name to see its information page, which will be similar to the one in **Figure 18-17**.

4. Tap the + button (subscribe button or, as Apple sometimes refers to it, follow button), in the upper-right corner.

   The podcast appears in the Library section of the app, and the newest episode is downloaded to your iPhone.

5. Tap Library in the toolbar at the bottom of the screen and then tap the name of the podcast you subscribed to and view its information screen.

6. Tap the more icon (three dots) and then tap Settings to see the settings for the podcast (see **Figure 18-18**). From here you can customize how the podcast downloads and organizes episodes. Tap Done in the upper-right corner when you're finished with the Settings options.

Tap to subscribe/follow this podcast

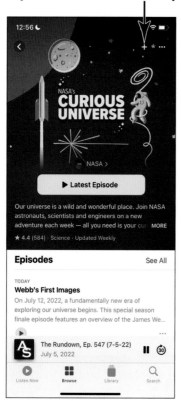

FIGURE 18-17

FIGURE 18-18

Play Podcasts

Playing podcasts is a breeze and works very much like playing audio files in the Music app.

1. Open the Podcasts app and tap Library at the bottom of the screen.

2. Tap the name of the podcast you'd like to listen to.

3. Tap the episode you want to play.

The episode begins playing; you can see the currently playing episode near the bottom of the screen, just above the toolbar.

4. Tap the currently playing episode to open the playback controls, as shown in **Figure 18-19**.

FIGURE 18-19

5. Drag the circle on the playhead in the middle of the screen to "scrub" to a different part of the episode, or tap the rewind or fast-forward icons to the left and right of the pause/play icon, respectively.

6. To adjust the playback speed, tap the 1x icon in the lower left of the playback controls and select another speed.

7. Adjust the volume by dragging the volume slider near the bottom of the screen, or by using the volume buttons on the side of your iPhone.

Tap the Listen Now button in the toolbar at the bottom left of the app's screen (refer to Figure 18-17) to see a list of the newest episodes that have been automatically downloaded to your iPhone.

IN THIS CHAPTER

» **Take pictures with your iPhone**

» **Save photos from the web**

» **View albums and single photos**

» **Edit and organize photos**

» **View photos by time and place**

» **Share and delete photos**

Chapter **19**

Taking and Sharing Photos

With its high-resolution camera modules and gorgeous screen, the iPhone is a natural for taking and viewing photos. It supports JPEG and HEIF (High Efficiency Image Format) photos. You can shoot your photos by using the iPhone cameras with built-in square or panorama mode. With recent iPhone models, you can edit your images using smart adjustment filters. You can also sync photos from your computer, save images you find online to your iPhone, or receive them by email, MMS, or iMessage.

The photo sharing feature lets you share groups of photos with people using iCloud on an iPhone or iPad or on a Mac or Windows computer with iCloud access. Your iCloud photo library makes all this storage and sharing easy.

When you have taken or downloaded photos to play with, the Photos app lets you organize them and view them in albums, one by one, or in a slideshow. You can view photos also by the years in which they were taken, with images divided into collections by the location or time you took them. With iOS 16, videos and live photos will play as you browse through Photos, making it a more dynamic and interesting experience. You can also AirDrop (iPhone 5 and later), email, message, or tweet a photo to a friend, print it, share it via AirPlay, or post it to Facebook. Finally, you can create time-lapse videos with the Camera app, allowing you to record a sequence in time, such as a flower opening as the sun warms it or your grandchild stirring from sleep. You read about all these features in this chapter.

Take Pictures with the iPhone Cameras

The cameras in the iPhone are top-notch, so you'll be pleased with the results no matter which phone model you have.

TIP For all iPhone models with a Home button, to open the camera when the lock screen is displayed, swipe up from the bottom of the screen and tap the Camera app icon in Control Center to go directly to the Camera app. If you have an iPhone model without a Home button, swipe down from the right corner of your screen to open Control Center and tap the Camera app icon in the lower right, or simply tap the Camera shortcut on the lock screen. You can also swipe down from the top of the screen to open Notification Center (with the iPhone unlocked) and then swipe from right to left near the top of the screen to access Camera. Another quick way to open the Camera app on either type of iPhone is to simply swipe the lock screen from right to left. Here are the basics step to follow in order to take a picture:

1. Tap the Camera app icon on the Home screen to open the app.

2. If the camera type at the bottom or side of the screen (depending on how you're holding your iPhone) is set to Video or something other than Photo, swipe to choose Photo (the still camera), as shown in **Figure 19-1**.

3. Make adjustments to the camera settings (discussed in the following paragraph), tap on the screen where you want the camera to focus (such as a person's face), and then tap the shutter button, which is the large white circle. You'll hear a camera shutter click as your iPhone captures the image.

Live photos   Switch cameras

Flash   Swipe these options to select Photo   Shutter

FIGURE 19-1

TIP

You can capture an image also by pressing the volume up or volume down button.

The following are adjustments you can make after you've opened the Camera app:

» To choose a still camera mode, select the Portrait, Pano (for panorama), or Square options by using the slider control above or next to the shutter button. These controls let you create flattering portraits or square images like those you see on Instagram. With Pano selected, tap to begin to take a picture, pan across the view, and then tap Done to capture the panoramic display.

» To set the flash, tap the flash icon (lightning bolt) when and then select a flash option:

• On, if your lighting is dim enough to require a flash

• Off, if you don't want iPhone to use a flash

• Auto, if you want to let iPhone decide for you

» To use the high dynamic range (HDR) feature, tap the HDR button to enable it. (If it's disabled, the button will appear with a line through it.) This feature uses several images, some underexposed and some overexposed, and combines the best ones into one image, sometimes providing a more finely detailed picture. If you don't see the HDR button, that means Smart HDR is already enabled in Settings ⇨ Camera. These settings don't apply to iPhone 13 models.

WARNING

The file size of HDR pictures can be very large, meaning they'll take up more of your iPhone's memory than standard pictures.

» If you want a time delay before the camera snaps the picture, tap the timer icon at the top of the screen, and then tap either 3s or 10s for a 3- or 10-second delay, respectively. If you don't see the timer icon, tap the small arrow in the top center of the screen, and then tap the timer icon, which is just above the shutter button (large white button).

» To take a live photo, tap the live icon (concentric circles). As opposed to freezing a single moment in time, the live photos feature lets you capture 3-second moving images, which can create some truly beautiful photos. Be sure to hold your iPhone still for at least 3 seconds so that you don't move too soon and cause part of your live photo to show the movement of your iPhone as you get into position for the picture.

» To apply color filters to your photos before taking them, tap the filters icon in the upper right if holding your iPhone in portrait mode. If holding it in landscape, tap the small arrow at the top center of the screen, and then tap the filters icon just above the shutter button. Swipe through the filter options above the shutter button to find one that suits you.

» Tap the area of the grid where you want the camera to autofocus.

TIP

A lot of people forget to tap on the screen to focus the lens, but doing so also helps the camera determine the proper exposure settings for your subject. So don't skip this small step when taking pictures.

» Place two fingers apart from each other on the screen and then pinch them together (still pressing down on the screen) to display a digital zoom control. Drag the circle in the zoom bar to the right or left to zoom in or out on the image.

» To switch between the front camera and rear camera, tap the switch camera icon. You can then take selfies (pictures of yourself), so go ahead and tap the shutter button to take another picture.

» To view the last photo taken, tap the thumbnail of the latest image in the bottom left in landscape mode or the bottom right in portrait mode; the Photos app opens and displays the photo.

After you're viewing the photo in the Photos app, you can take the following actions:

» Tap the share icon (box with an arrow, located in the bottom-left corner of the screen) to open the share sheet, displaying a row of icons that allow you to AirDrop, email, or instant message the photo, assign it to a contact, use it as iPhone wallpaper, tweet it, post it to Facebook, share via iCloud Photo Sharing or other apps you have installed, or print it (see **Figure 19-2**). You can tap images to select more than one.

» To delete a displayed image, tap the trash icon, then tap Delete Photo in the confirmation menu that appears.

TIP

You can use the iCloud photo sharing feature to automatically sync your photos across various devices. Turn on iCloud photo sharing by tapping Settings on the Home screen, tapping Photos, and then toggling the iCloud Photos switch on (green).

FIGURE 19-2

Save Photos from the Web

The web offers a wealth of images that you can download to your Photos library. To save an image from the web, follow these steps:

1. Open Safari and navigate to the web page containing the image you want, as shown in **Figure 19-3**.

TIP

For more about how to use Safari to navigate to or search for web content, see Chapter 12.

2. Touch and hold down on the image.

A menu appears on the screen, as shown in **Figure 19-4**.

3. Tap Add to Photos.

The image is saved to your Recents album in the Photos app.

FIGURE 19-3　　　　　　　**FIGURE 19-4**

TIP

If you want to capture your iPhone screen as a photo, the process is simple. Press the sleep/wake button or the side button (depending on your iPhone model) and Home button simultaneously. (Press the side and volume up buttons for iPhone models without a Home button.) When the screen flashes white, the screen capture is complete. The capture is saved in PNG format to your Recents album in the Photos app.

View an Album and Its Photos

The Photos app organizes your pictures into albums. It has a default album called Recents, which stores photos taken with your iPhone's camera. It may also contain other folders created when you synced photos on your computer to your iPhone, or perhaps if you synced from other devices through iTunes or shared via Photos (from your iPad, for instance).

TIP

The Hidden album and the Recently Deleted album are both locked by default in iOS 16. They can be unlocked using only Face ID or Touch ID (depending on your iPhone model) or your iPhone's passcode.

1. To view your albums, start by tapping the Photos app icon on the Home screen.

2. Tap the Albums icon at the bottom of the screen to display your albums, as shown in **Figure 19-5**.

3. Tap an album. The photos in it are displayed.

TIP

You can associate photos with faces and events. When you do, additional tabs appear at the bottom of the screen when you display an album containing that type of photo.

You can view photos individually by opening them from an album.

1. Tap the Photos app icon on the Home screen.

2. Tap Albums (refer to Figure 19-5).

3. Tap an album to open it. Then, to view a photo, tap it.

   The picture expands, as shown in **Figure 19-6**.

4. Flick your finger to the left or right to scroll through the album to look at the individual photos in it.

5. To return to album view, tap the back icon (left-pointing arrow) in the upper-left corner and then the Albums button.

TIP

You can place a photo on a person's information record in Contacts. For more on how to do this, see Chapter 7.

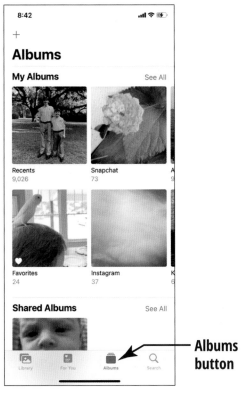

Albums button

FIGURE 19-5

FIGURE 19-6

Edit Photos

The Photos app isn't Photoshop, but it does provide some tools for editing photos.

1. Tap the Photos app icon on the Home screen to open the app.

2. Using methods previously described in this chapter, locate and display a photo you want to edit.

3. Tap the Edit button in the upper-right corner of the screen.

   The Edit Photo screen appears with various tools. The one shown in **Figure 19-7** is for a photo shot in portrait mode.

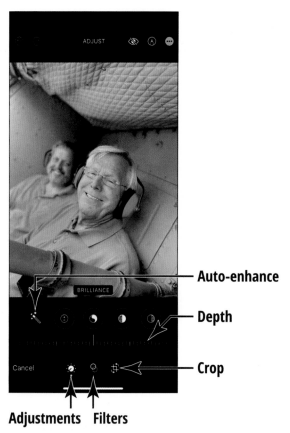

Auto-enhance

Depth

Crop

Adjustments Filters

FIGURE 19-7

4. At this point, you can take several possible actions with these tools (their screen location depends on whether you're holding your iPhone in landscape mode or portrait mode):

- *Auto-enhance:* The icon for this feature looks like a magic wand, and it pretty much works like one. Tapping the wand allows your iPhone to apply automatic adjustments to your photo's exposure, saturation, contract, and more.

- *Adjustments:* Swipe the options just above the Depth slider and immediately below the photo to see adjustment options such as Light, Color, and B&W. You can use a slew of other tools to tweak contrast, color intensity, shadows, and more.

- *Filters:* Apply any of nine filters (such as Vivid, Mono, or Noir) to change the look of your image. These effects adjust the brightness of your image or apply a black-and-white tone to your color photos. Tap the filters icon in the middle of the tools at the bottom of the screen and scroll to view available filters. Tap one and then tap Done to apply the effect to your image.

If you want to revert to the original image and remove previously applied filters, use the Original filter.

- *Crop:* To crop the photo to a portion of its original area, tap the crop icon. You can then tap any corner of the image and drag to remove areas of the photo. Tap Crop and then Save to apply your changes.

- *Depth:* With this slider, you can control the depth of field of any photo shot in portrait mode. A lower value, such as *f*1.8, gives you more background blur. A higher number, such as *f*16, reveals a more detailed background.

Each of the editing features has a Cancel button. If you don't like the changes you've made, tap this button to stop making changes before you save the image.

5. If you're pleased with your edits, tap the Done button. A copy of the edited photo is saved.

Organize Photos

You'll probably want to organize your photos to make it simpler to find what you're looking for.

1. If you want to create your own album, open the Recents album.

2. Tap the Select button in the top-right corner and then tap individual photos to select them.

Small check marks appear on the selected photos, as shown in **Figure 19-8**.

FIGURE 19-8

3. Tap the share icon (square with an escaping arrow) in the lower-left corner, tap Add to Album, and then tap New Album.

If you've already created albums, you can choose to add the photo to an existing album at this point. Tap the share icon (square with an escaping arrow) in the lower-left corner, tap Add to Album, and then tap the album that you want to add the photo to.

4. Enter a name for a new album and then tap Save.

If you've created a new album, it appears in the Photos main screen with the other albums.

When you've selected photos in Step 2, you can choose several other share options or tap Delete. This allows you to share or delete multiple photos at a time.

Share Photos with Mail, Twitter, or Facebook

You can easily share photos stored on your iPhone by sending them as email attachments or as a text message, by posting them to Facebook, by sharing them via iCloud photo sharing or Flickr, or as tweets on Twitter. (See Chapter 14 for information on installing third-party apps.)

1. Tap the Photos app icon on the Home screen.

2. Tap the Library or Albums button and locate the photo you want to share.

3. Tap the photo to select it and then tap the share icon (box with an arrow jumping out of it). When the share sheet shown in **Figure 19-9** appears, you can tap additional photos to select them.

4. Tap the AirDrop, Messages, Mail, Facebook, Twitter, Flickr, or any other option you'd like to use.

5. In the message form that appears, make any modifications that apply in the To, Cc/Bcc, and Subject fields and then type a message for email or enter your Facebook posting or Twitter tweet.

6. Tap the Send button or Post button, and the message and photo are sent or posted.

FIGURE 19-9

TIP

You can also copy and paste a photo into documents, such as those created in the Pages word-processor app. To do this, tap a photo in Photos and then tap the share icon. Tap the Copy command. In the destination app, press and hold down on the screen and tap Paste.

Share a Photo Using AirDrop

AirDrop, available to users of iPhone 5 and later, provides a way to share content, such as photos, with others who are nearby and who have an AirDrop-enabled device (more recent Macs that can run OS X 10.10 or later).

Follow the steps in the previous task to locate a photo you want to share.

1. Tap the share icon.

2. If an AirDrop-enabled device is in your immediate vicinity (within 30 feet or so), you see the device listed (see **Figure 19-10**). Tap the device name and your photo is sent to the other device.

TIP

Other Apple devices (iPhones or iPads) must have AirDrop enabled to use this feature. To enable AirDrop, open Control Center (swipe up from the bottom of any screen, or down from the upper-right corner for iPhone models without a Home button as well as iPads) and tap AirDrop. If you don't see AirDrop as an option, press and hold down on the Communications area in Control Center. The Communications area houses Wi-Fi and other options; the options will expand and you'll see AirDrop there, as shown in **Figure 19-11**. Choose Contacts Only or Everyone to specify whom you can use AirDrop with.

AirDrop-enabled devices nearby

FIGURE 19-10

Tap AirDrop

FIGURE 19-11

Share Photos Using iCloud Photo Sharing

iCloud photo sharing allows you to automatically share photos using your iCloud account.

1. Select a photo or photos you would like to share, and tap the share icon in the lower left.

2. In the share sheet that opens, tap Add to Shared Album (refer to Figure 19-9).

3. Enter a comment if you like, tap Shared Album and choose the album you want to use.

4. Tap iCloud to go back, and then tap Post.

   The photos are posted to your iCloud Photo Library.

Delete Photos

You might find that it's time to get rid of some of those old photos of the family reunion or the latest community center project. If the photos weren't transferred from your computer, but instead were taken, downloaded, or captured as screenshots on the iPhone, you can delete them.

1. Tap the Photos app icon on the Home screen.

2. Tap Albums and then tap an album to open it.

3. Tap a photo that you want to delete, and then tap the trash icon. In the confirming dialog that appears, tap the Delete Photo button to finish the deletion.

WARNING

If you delete a photo in photo sharing, it's deleted on all devices that you shared it with.

TIP

If you'd like to recover a photo you've deleted, tap Albums, scroll to the bottom of the page, and tap Recently Deleted. Tap the photo you want to retrieve, tap the Recover button in the lower-right corner, and then tap the Recover Photo button when prompted.

Chapter **20**

Creating and Watching Videos

All current and recent iPhone models sport both front and rear video cameras that you can use to capture your own videos, which iOS enables you to edit in the same way you edit photos: You can apply adjustments and filters as well as crop your videos. Speaking of editing, you can also download the iMovie app for iPhone (a more limited version of the longtime mainstay on Mac computers) and do an editing deep dive by adding titles, music, transitions, and much more.

Using the TV app (formerly known as Videos), you can watch downloaded movies or TV shows, as well as media you've synced from iCloud on your Mac or PC, and even media provided from other content providers, such as cable and streaming video services. The TV app aims to be your one-stop shop for your viewing pleasure.

In this chapter, I explain all about shooting and watching video content from a variety of sources. For practice, you might want to refer to Chapter 16 first to find out how to purchase or download one of many available TV shows or movies from the iTunes Store.

Capture Your Own Videos with the Built-In Cameras

The camera lenses on newer iPhones have perks for photographers, including large apertures and highly accurate sensors, which make for better images all around. In addition, auto image stabilization makes up for any shakiness in the hands holding the phone, and autofocus has sped up thanks to super-fast processors. Videographers will appreciate the fast frames-per-second capability as well as the slow-motion feature.

1. To capture a video, tap the Camera app icon on the Home screen.

   On your iPhone, two video cameras are available for capturing video, one from the front and one from the back of the device. (See more about this topic in the next task.)

   The Camera app opens, as shown in **Figure 20-1**.

2. Tap and slide the camera-type options above or next to the red record button (depending on how you're holding your iPhone) until Video rests above or next to the button.

   This is how you switch from the still camera (photo mode) to the video camera (video mode).

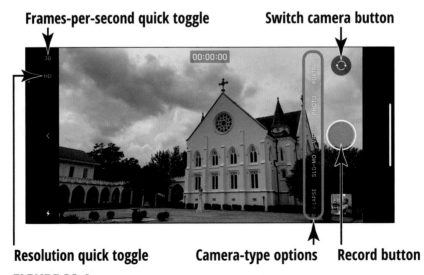

Frames-per-second quick toggle **Switch camera button**

Resolution quick toggle **Camera-type options** **Record button**

FIGURE 20-1

3. If you want to switch between the front and back cameras, tap the switch camera icon in the top-right corner of the screen when holding your iPhone in landscape mode (refer to Figure 20-1) or in the bottom-right when holding the iPhone in portrait mode.

4. Use the quick toggle icons in the upper left of your screen (landscape mode) or upper right (portrait mode) to easily change the resolution and frame rates for your video. Tap HD or 4K to switch between these modes, and tap the number to switch between 30 and 60 frames per second in HD or 24, 30, and possibly 60 (depending on your iPhone model) frames per second in 4K.

Quick toggles automatically appear with iPhone 14 models, iPhone 13 models, iPhone 12 models, the iPhone SE (second generation), iPhone 11 models, iPhone XS models, and the iPhone XR. For other models, go to Settings ⇨ Camera ⇨ Record Video and enable Video Format Control.

5. To begin recording the video, tap the red record button.

The red circle in the middle of this button turns into a red square when the camera is recording.

6. To stop recording, tap the record button again.

Your new video is displayed in the bottom-left or bottom-right corner of the screen, depending on how you're holding your iPhone. Tap the video to play, share, or delete it. In the future, you can find and play the video in your Recents album when you open the Photos app.

Using QuickTake for Videos

QuickTake videos are videos you can record in photo mode. So, if you're taking a picture of someone but feel like recording part of the action, use QuickTake.

1. Open the Camera app.

2. While in photo (still picture) mode, press and hold down on the white shutter button.

The shutter button turns red and the recording begins.

3. To stop recording, simply remove your finger from the screen.

If in Step 2 you'd like to continue recording hands-free, press and hold down on the shutter button. When recording begins, slide the record button to the right to lock it in recording mode, as shown in **Figure 20-2**. You can still take photos while locked in recording mode by tapping the white shutter button to the right. Tap the record button to stop recording and return to photo mode.

QuickTake Photo Shutter button

QuickTake Video Record button

FIGURE 20-2

QuickTake is supported on iPhone 14 models, iPhone 13 models, iPhone 12 models, the iPhone SE (second generation), iPhone 11 models, iPhone XS models, and the iPhone XR.

TIP

Before you start recording, remember where the camera lens is — while holding the iPhone and panning, it's easy to mistakenly put your fingers directly over the lens! Also, you can't pause your recording; when you stop, your video is saved, and when you start recording, you're creating a new video file.

Edit Videos

The Photos app (where your videos are stored) isn't Final Cut Pro, but it does provide a few handy tools for editing videos.

1. Tap the Photos app icon on the Home screen, locate your video, and tap it to open it.

2. Tap the Edit button in the upper-right corner of the screen.

The Edit screen appears. The one shown in **Figure 20-3** is for a video shot in portrait mode (even though the iPhone is held in landscape orientation).

FIGURE 20-3

3. At this point, you can take several possible actions with the tools provided. Tap the icon for the following tools to access their sets of options:

- *Crop/rotate:* To rotate or flip the image, tap the rotate or flip icon in the upper-left corner (next to Cancel when viewing in portrait mode). To crop the video to a portion of its original area, tap the crop/rotate icon. You can then tap any corner of the image and drag to remove areas of the video. Tap Crop and then Save to apply your changes.

- *Filters:* Apply one of nine filters (such as Vivid, Mono, or Noir) to change the look of your video images. These effects adjust the brightness of your video or apply a black-and-white tone to your color videos. Tap the filters icon and then scroll through the list to view available filters. Tap one to apply the effect to your video.

- *Adjustments:* Tap Light, Color, or B&W to access a slew of tools that you can use to tweak contrast, color intensity, shadows, and more.

- *Auto-enhance:* The icon for this feature, which looks like a magic wand, appears when you tap the adjustments icon. Tapping the wand allows your iPhone to apply automatic adjustments to your video's exposure, saturation, contrast, and so on.

- *Trim:* Use the trim tool to remove parts of your video you no longer want to view.

4. If you're pleased with your edits, tap the Done button.

A copy of the edited video is saved.

Each of the editing features has a Cancel button. If you don't like the changes you made, tap this button to stop making changes before you save the image. What if you make changes you later regret? Just open the video, tap Edit, and then tap the red Revert button in the upper right to discard changes to the original.

Play Movies or TV Shows with the TV App

Open the TV app for the first time and you'll be greeted with a Welcome screen; tap Get Started. You'll be asked to sign in to your television provider, if you haven't done so already.

Signing in allows you to use the TV app to access content in other apps (like ESPN or Disney), if such services are supported by your TV provider. This way, you need to use only the TV app — as opposed to using multiple apps — to access content and sign into the services.

Should you decide to skip signing in to your TV provider and worry about it later (or if you've already opened the TV app and cruised right past this part), you can access the same options by going to Settings ⇨ TV Provider (see **Figure 20-4**), tapping the name of your provider, and then entering your account information.

If you're not sure of your account information, you'll need to contact your provider for assistance.

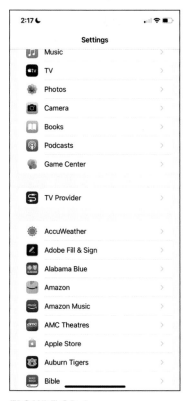

FIGURE 20-4

The TV app offers a couple of ways to view movies and TV shows: via third-party providers (many require a subscription, such as Amazon's Prime Video, Disney+, and Netflix) or items you've purchased or rented from the iTunes Store.

Content from third-party providers

To access content from providers like Apple TV+, NBC, ABC, and PBS, open the TV app and tap the Watch Now icon in the bottom left of your screen (see **Figure 20-5**). Swipe to see hit shows and browse by genres such as Comedy or Action.

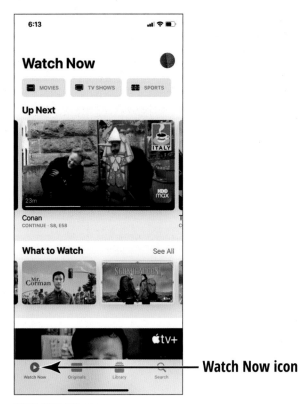

Watch Now icon

FIGURE 20-5

Tap a show that interests you, like I did in **Figure 20-6**. Tap an episode to begin playing it, or tap the Details button (you may need to swipe down a bit to find it) to see a description of the episode, as shown in **Figure 20-7**. If you have the app that supports the video, the video will open automatically in the correct app. You may be prompted to connect apps from providers like PBS and ABC to the TV app so you can watch their videos in TV. If you want to do so, tap Connect; if not, just tap Not Now. If you don't have the app installed that you need to watch the video, you'll be asked if you'd like to download and install it.

TIP

With Home Sharing set up for your iPhone and computer, you can stream videos from your computer to your iPhone. Check out this Apple Support article for more info regarding Home Sharing and how to set it up for your devices: `https://support.apple.com/en-us/HT202190`.

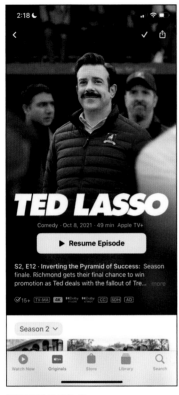

FIGURE 20-6

FIGURE 20-7

Content from the iTunes Store

To access video you've purchased or rented from the iTunes Store, follow these steps:

1. Tap the TV app icon on the Home screen to open the application, and then tap Library at the bottom of the screen.

   A screen similar to the one in **Figure 20-8** appears.

2. Tap the appropriate category at the top of the screen (TV Shows, Movies, Family Sharing, Genres, and more, depending on the content you've downloaded), and then tap the video you want to watch.

   Information about the movie or TV show episodes appears.

3. For a TV show, tap the episode that you'd like to play. For a movie, tap the play icon (which appears on the description screen).

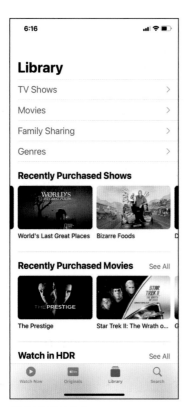

FIGURE 20-8

The movie or TV show begins playing. If you see a small, cloud-shaped icon instead of a play icon, tap it and the content is downloaded from iCloud.

4. With the playback tools displayed (as shown in **Figure 20-9**), take any of these actions:

- To pause playback, tap the pause icon.

- To move to a different location in the video playback, drag the video slider or tap one of the buttons labeled 10 to go backward or forward 10 seconds.

  If a video has chapter support, a Scenes button appears for displaying all chapters so that you can move more easily from one to another. You can also tap Go to Previous Chapter or Go to Next Chapter to navigate.

- To decrease or increase the volume, tap the end of the lighter section of the volume slider and drag it left or right, respectively.

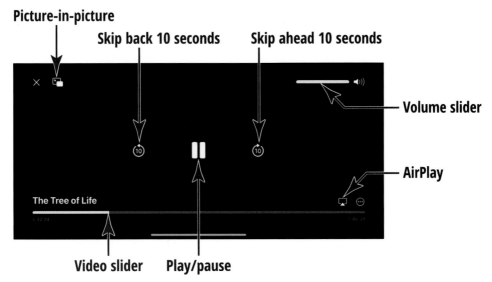

Picture-in-picture

Skip back 10 seconds

Skip ahead 10 seconds

Volume slider

AirPlay

The Tree of Life

Video slider   **Play/pause**

FIGURE 20-9

TIP

If your controls disappear during playback, just tap the screen and they'll reappear.

5. To stop the video and return to the information screen, tap the Done button to the left of the progress bar.

Turn On Closed-Captioning

The iTunes Store and iPhone offer support for closed-captioning and subtitles. To use this feature, look for the CC logo on media that you download.

TECHNICAL
STUFF

Video that you record doesn't have this capability.

If a movie has either closed-captioning or subtitles, you can turn on the feature on your iPhone:

1. Tap the Settings icon on the Home screen.

2. Tap Accessibility, and then scroll down and tap Subtitles & Captioning in the Hearing section.

3. On the menu that appears, tap the Closed Captions + SDH switch (see **Figure 20-10**) to turn the feature on (green).

FIGURE 20-10

Now when you play a movie with closed-captioning, you can tap the more icon (three dots) to the right of the playback controls, and then tap Subtitles to manage these features.

Delete a Video from the iPhone

You can buy videos directly from your iPhone, or you can sync via iCloud, Finder (a Mac running macOS Catalina or newer), or iTunes (a PC or a Mac running macOS Mojave or earlier) to place content you've bought or created on another device on your iPhone.

When you want to get rid of memory-hogging video content on your iPhone, do the following:

1. Open the TV app and then go to the TV show or movie you want to delete.

2. Tap the downloaded icon to the right of the title (blue circle with a check mark).

3. Tap Remove Download in the confirmation dialog that appears.

   The downloaded video will be deleted from your iPhone.

If you buy a video from Apple on your computer, sync to download it to your iPhone, and then delete it from your iPhone, it's still saved in your Library. You can sync your computer and iPhone again to download the video. Remember, however, that rented movies, when deleted, are gone with the wind. Also, video — unlike photos and music — doesn't sync to iCloud.

IN THIS CHAPTER

» **View your current location**

» **Change views and zoom**

» **Go to other locations or favorites**

» **Drop markers and find directions**

» **Receive turn-by-turn navigation help**

» **Get a bird's-eye view of the world**

Chapter **21**

Navigating with Maps

The Maps app has lots of useful functions. You can find directions with alternative routes from one location to another. You can bookmark locations to return to them. And the Maps app delivers information about locations, such as phone numbers and web links to businesses. You can even add a location to your Contacts list or share a location link with your buddy using Mail, Messages, Twitter, or Facebook. The Nearby feature helps you explore local attractions and businesses, and transit view lets you see public transit maps for select cities around the world.

Maps also includes other great features, such as lane guidance (helps you get in the correct lane for upcoming turns), speed limit indicators for many roads, and maps of such indoor locations as shopping centers and airports. You can also view the balance and replenish funds of transit cards. With the Look Around feature, Maps will alert you to speed cameras and red-light cameras, and will help owners of electric vehicles plan their routes based on the availability of charging stations. iOS 16 even includes a great view of the entire globe that you can use to discover the world!

You're about to have some fun exploring Maps in this chapter. Just don't do it while driving!

TIP

Some features are available only in certain locations. To see what's available in your neck of the woods, check out the Maps area of this site: `www.apple.com/ios/feature-availability`.

Display Your Current Location

iPhone can figure out where you are at any time and display your current location.

1. From the Home screen, tap the Maps icon. Tap the current location button (labeled in **Figure 21-1**).

A map is displayed with your current location indicated with a marker.

TECHNICAL STUFF

Depending on your connection, Wi-Fi or cellular, a pulsating circle may appear around the marker, indicating that the area surrounding your location is based on cell tower triangulation. Your exact location can be anywhere within the area of the circle, and the location service is likely to be less accurate using a Wi-Fi connection than your phone's cellular connection.

2. Double-tap the screen to zoom in on your location.

(Additional methods of zooming in and out are covered in the "Zoom In and Out" section in this chapter.)

TIP

As mentioned, if you access maps via a Wi-Fi connection, your current location is a rough estimate based on triangulation. Your iPhone can more accurately pinpoint where you are by using your cellular data connection, your iPhone's global positioning system (GPS), and Bluetooth. But you can get pretty accurate results with just a Wi-Fi–connected iPhone if you type a starting location and an ending location to get directions.

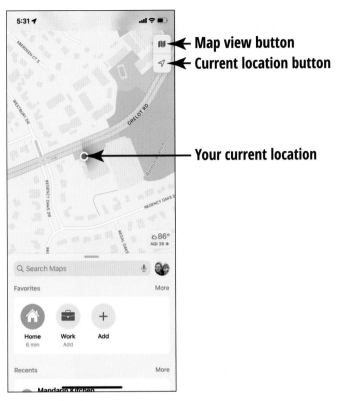

Map view button ◄—
Current location button ◄—

Your current location ——

FIGURE 21-1

Change Views

The Maps app offers four primary views: explore, driving, transit, and satellite. iPhone displays the explore view by default the first time you open Maps. To change views:

1. With Maps open, tap the map view (aka map modes) button in the upper right of the screen (refer to Figure 21-1) to reveal the Choose Map dialog, as shown in **Figure 21-2.** The icon reflects the currently selected map view.

2. Tap the Transit button, and then tap the X button (close) in the upper right of the dialog.

In transit view, you see data about public transit if you're viewing a major city for which transit information is available.

3. Tap the map view button again, tap the Satellite button, and then tap the X button.

   The satellite view appears, as shown in **Figure 21-3**.

4. You can display a 3D effect (see **Figure 21-4**) for any view by swiping up on the screen with two fingers (or simply tap the 3D button in satellite view). Tap the 2D button, which replaces the 3D button, to revert to two-dimensional imaging.

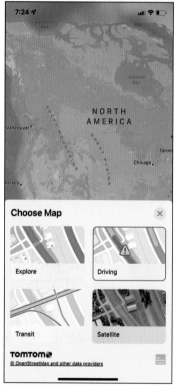

FIGURE 21-2

FIGURE 21-3

TIP

Maps displays a weather icon in the lower-right corner of a map to indicate the weather conditions in the area. Press and hold down on the weather icon to display even more weather-related info.

FIGURE 21-4

Zoom In and Out

You'll appreciate the zoom feature because it gives you the capability to zoom in and out to see more detailed or less detailed maps and to move around a displayed map.

1. With a map displayed, double-tap with a single finger to zoom in, as shown in **Figure 21-5**.

The image on the left shows the map before zooming in, and the image on the right shows the map after zooming and using 3D.

2. Double-tap with two fingers to zoom out, revealing less detail.

3. Place two fingers together on the screen and then move them apart to zoom in.

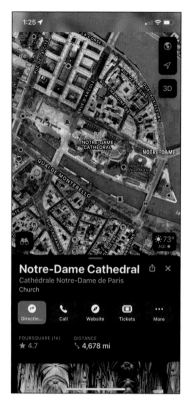
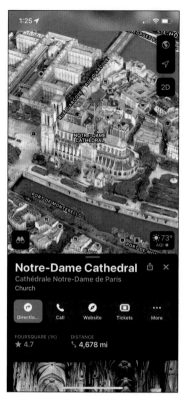

FIGURE 21-5

4. Place two fingers apart on the screen and then pinch them together to zoom out.

5. Double-tap the screen, but on the second tap hold your finger down on the screen. Now, drag your finger up or down the screen to zoom in or out.

This action is called one-handed zoom, and it allows you to zoom in and out with just one finger.

6. Drag the map in any direction to move to an adjacent area.

It can take a few moments for the map to redraw itself when you enlarge, reduce, or move around it, so be patient. Areas that are being redrawn look like blank grids but are filled in eventually. Also, if you're using satellite view, zooming in may take some time; wait it out. The blurred image will resolve and looks quite nice, even tightly zoomed.

Go to Another Location or a Favorite

You don't have to view just your current location. Maps allows you to view cities across the globe by simply typing their name in the search field.

1. With Maps open, tap in the search field to display the keyboard, as shown in **Figure 21-6**.

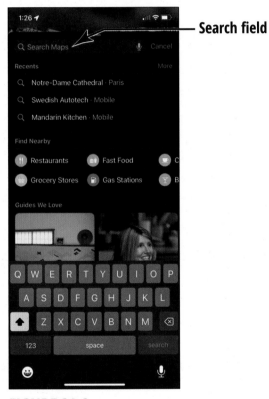

— **Search field**

FIGURE 21-6

2. Type a location, using either a street address with city and state, a stored contact name, or a destination (such as Empire State Building or Lincoln Memorial).

Maps may make suggestions as you type if it finds any logical matches. Tap the result you prefer. The location appears with a marker on it (the marker depends on the type of establishment, such as restaurant or landmark) and an information dialog with the location name and a blue travel mode/time button (see **Figure 21-7**). The icon in the blue travel mode/time button indicates the mode of travel, such as driving or transit.

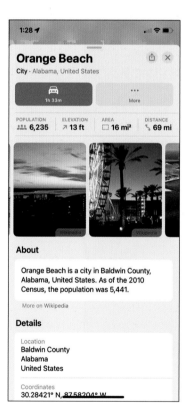

FIGURE 21-7

3. Swipe up on the information dialog and more information is displayed, such as pictures of the location, address, and phone number.

4. To move back to the map, swipe down on the information dialog. To move to a nearby location, tap the screen and drag in any direction.

Drop a Marker (or Pin)

Markers and pins are the same, but the iOS 16 version of Maps leans toward using markers as the default term. I felt that this part of the chapter was the best place to explain that difference, in case you've used iOS devices in the past and have always referred to marked locations as pins and were wondering what this marker business was all about.

1. Display a map that contains a spot where you want to drop a marker to help you find directions to or from that site.

   If you need to, you can zoom in to a more detailed map to see a better view of the location you want to mark.

2. Press and hold down on the screen at the location where you want to place the marker.

   The marker appears, together with an information dialog, as shown in **Figure 21-8**.

3. Swipe up on the information dialog to display details about the marker location, as shown in **Figure 21-9**.

TIP

If a location has associated reviews on sites such as the restaurant and travel review site Yelp (www.yelp.com), you can display details about the location and scroll down to read the reviews.

FIGURE 21-8

FIGURE 21-9

Find Directions

You can get directions in a couple of different ways.

1. Tap a marker on your map and then tap the blue travel mode/time button in the information dialog.

   A blue line appears, showing the route between your current location and the chosen marker, as shown in **Figure 21-10**. Sometimes alternate routes will appear as well (in a lighter shade of blue), allowing you to select which you'd rather take; just tap the alternate route if you deem it's best for you.

2. Tap the green Go button to get started, or tap the close icon (X) on the right side to return to the Maps main screen.

You can also manually enter two locations to get directions from one to the other:

1. To enter two locations to get directions from one to the other, tap the search field to open the keyboard. Begin to type the location you'd like to get to, and then select the location from the list.

In the information dialog that appears, you see the starting point at the top and the first (and maybe only) stop just below it (shown in **Figure 21-11**).

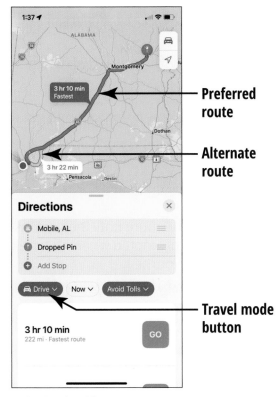

Preferred route

Alternate route

Travel mode button

FIGURE 21-10

FIGURE 21-11

2. My Location is the default, but if you want to change it to a different location, tap My Location, type the new starting point, and then tap Route in the upper-right corner.

TIP

At the bottom of the directions dialog (refer to Figure 21-11) you see the default drive icon, which you can tap to also see the walk, transit, cycling, and ride icons. Tap one of them to see directions optimized for that mode of transportation.

3. Tap the green Go button when you're ready to start on your route.

Get Turn-by-Turn Navigation Help

Maps is all too happy to help you with voice instructions to move from place to place.

1. After you've found directions, tap the green Go button to get started.

The narration begins and text instructions are displayed at the top of the screen, as shown in **Figure 21-12**. Estimated arrival time and distance are shown at the bottom of the screen in the collapsed information dialog.

2. Continue on your way according to the route until the next instruction is spoken. If necessary, you'll be prompted when it's time to switch lanes at various points on the journey.

3. To share an ETA with others at any time, swipe up on the information dialog, and then tap the Share ETA button that appears. Select a contact from the list of your favorites (as shown in **Figure 21-13**), or tap Open Contacts to select the one you want to share with, and then go along your merry way.

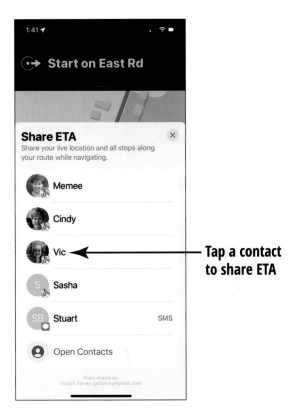

Tap a contact
to share ETA

FIGURE 21-12   **FIGURE 21-13**

At the bottom of the screen, you can see that you're sharing your ETA with the contact you've selected. To stop sharing, swipe up on the information dialog. Tap the Sharing with *x* Person/People button (where *x* is the number of people), and tap the contact. Then swipe down to return to your directions.

4. To see the details of your route, as shown in **Figure 21-14**, tap the top of the screen. Swipe up to resume turn-by-turn navigation.

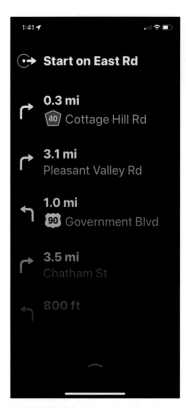

FIGURE 21-14

5. To turn the spoken navigation aid on or off, tap the audio icon on the right. Adjust the audio settings to your liking (as shown in **Figure 21-15**).

6. Maps ends the route automatically after you arrive, but you can end the route manually by swiping up on the information dialog and tapping the End Route button, as shown in **Figure 21-16**.

FIGURE 21-15

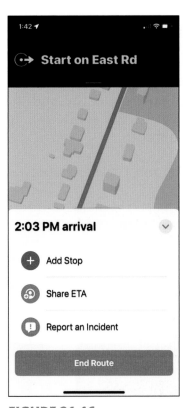

FIGURE 21-16

TIP

It's no problem if you need to perform another task while in the middle of getting directions. Press the Home button or swipe up from the bottom of the screen (for iPhone models without a Home button) to get out of Maps. A blue bar appears at the top of your screen. After your other task is complete, simply tap that blue bar to get back to your navigation in Maps.

Go Globetrotting!

I love the interactive 3D globe in Maps for iOS 16! You can tour the world and check out the sights, right from your iPhone. It kind of reminds me of the globe my teacher used to have in our fourth-grade homeroom.

To view the interactive globe:

1. Zoom out and continue to do so until you see the globe of the Earth, as shown in **Figure 21-17**.

2. Swipe the screen to rotate the globe, and pinch to zoom in and out of areas, as I'm doing in **Figure 21-18**.

3. Have fun!

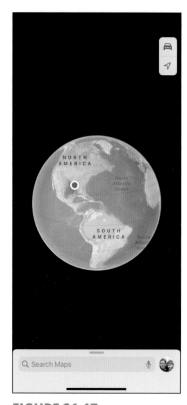

FIGURE 21-17

FIGURE 21-18

5

Living with Your iPhone

IN THIS PART . . .

Scheduling your life

Receiving reminders and notifications

Keeping track of your health

Troubleshooting and maintaining your iPhone

IN THIS CHAPTER

» Add calendar events

» Create a repeating event

» View, search for, and delete events

» Add an alert to an event

» Display clocks

» Set and delete alarms

» Use the Stopwatch and Timer apps

Chapter **22**

Keeping on Schedule with Calendar and Clock

Whether you're retired or still working, you have a busy life full of activities (perhaps even busier if you're retired, for some unfathomable reason). You need a way to keep on top of all those activities and appointments. The Calendar app on your iPhone is a simple, elegant, electronic daybook that helps you do just that.

In addition to being able to enter events and view them in a list or by the day, week, or month, you can set up Calendar to send alerts to remind you of your obligations and search for events by keywords. You can even set up repeating events, such as birthdays, monthly get-togethers with friends, or weekly babysitting appointments with the kids in your life. To help you coordinate calendars on multiple devices, you can also sync events with other calendar accounts. And by taking advantage of the Family Sharing feature, you can create a Family calendar that everybody in your family can view and add events to.

Another preinstalled app that can help you stay on schedule is Clock. Though simple to use, Clock helps you view the time in multiple locations, set alarms, check yourself with a stopwatch feature, and use a timer.

In this chapter, you master the simple procedures for getting around your calendar, creating a Family calendar, entering and editing events, setting up alerts, syncing, and searching. You also learn the straightforward ins and outs of using Clock.

View Your Calendar

Calendar offers several ways to view your schedule.

1. Start by tapping the Calendar app icon on the Home screen to open the app.

   Depending on what you last had open and the orientation in which you're holding your iPhone, you may see today's calendar, list view, the year, the month, the week, an open event, or the Search screen with search results displayed.

   TIP If you happen to own an iPhone model with a wider screen, when you hold it horizontally (landscape) in the Calendar app, you see more information on the screen than you see on other iPhone models. For example, in month view, you see the entire month on the left and detailed information on events for the selected day.

2. Tap Today at the bottom of the screen to display the current date in day view (if it isn't already displayed), and then tap the list view icon to see all scheduled events for that day.

   List view, shown in **Figure 22-1**, displays your daily appointments for every day in a list, with times listed on the right. Tap an event in the list to get more event details.

FIGURE 22-1

3. To display events only from a particular calendar or set of calendars, such as the Birthday or US Holidays calendars, tap Calendars at the bottom of list view and select which calendar(s) to view by tapping the circle to the left of the calendar name(s). Tap Done in the upper-right when finished.

A check mark in the circle indicates that calendar's items will be displayed.

TIP

You can hide or show all calendars for individual accounts in one fell swoop by tapping Hide All or Show All in the upper-right corner of each account.

4. Tap the current month in the upper-left corner to display months; in month view, the list view icon changes to a combined view icon. Tap this to show the month calendar at the top and a scrollable list of events at the bottom, as shown in **Figure 22-2**. Tap the icon again to return to month view.

Combined view

FIGURE 22-2

TECHNICAL
STUFF

5. Tap a date in month view; you see the week containing that date at the top and the selected day's events below.

The week view can't display combined lists.

6. In month view, note the year displayed at the top left of the screen. Tap the arrow to its left to get a yearly display (see **Figure 22-3**), and then tap a month to display it.

In year view, you see the calendar for the entire year with the current day circled in red.

You can move from one month to the next in month view also by scrolling up or down the screen.

7. To jump back to today view, tap Today in the bottom-left corner of the screen. The month containing the current day is displayed.

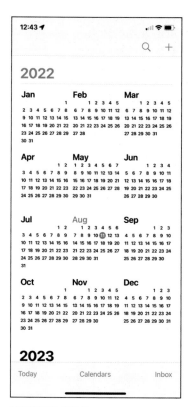

FIGURE 22-3

TIP

To view any invitation that you accepted, which placed an event on your calendar, tap Inbox in the lower-right corner. A list of accepted invitations is displayed. To view them all, tap the New tab and then the Replied tab. Tap Done to return to the calendar.

Add Calendar Events

Calendars are fun, but adding events to them makes them functional (my apologies; couldn't help it). Here's how to add events to calendars:

1. With any view displayed, tap + (add) in the upper-right corner of the screen to add an event.

    The New Event dialog appears.

2. Enter a title for the event and, if you want, a location.

3. Tap the All-Day switch to turn it on for an all-day event, or tap the Starts and Ends fields to set start and end times for the event.

As shown in **Figure 22-4**, a calendar or a clock appears when you tap the date or time button, respectively.

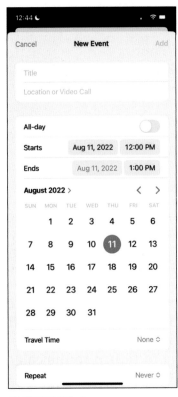

FIGURE 22-4

4. To select which calendar to use for the event, add a note, or change other settings, scroll down in the New Event dialog and do the following:

   • To select a different calendar, tap the Calendar button, tap the calendar you'd like to use for this event, and then tap New Event in the upper left.

- To add a note, tap in the Notes field, type your note, and then tap the Add button to save the event.

- To change another setting, tap its button, tap any options you'd like to adjust, and then tap New Event in the upper-left corner.

You can edit any event at any time. Simply tap the event in any view of your calendar and, when the details are displayed, tap Edit in the upper-right corner. The Edit Event dialog appears, offering the same settings as the New Event dialog. Tap the Done button in the upper right to save your changes or Cancel in the upper left to return to the calendar without saving any changes.

Add Events with Siri

Time to play around with Siri and the Calendar app; it's a lot of fun!

1. Press and hold down on the Home button (side button for iPhone models without a Home button) or say "Hey Siri."

2. Speak a command, such as "Hey Siri. Create a meeting on October 3rd at 2:30 p.m."

The event is automatically added to Calendar.

You can schedule an event using Siri in several ways:

» Say something like "Create event." Siri asks you first for a date and then for a time.

» Say something like "I have a meeting with John on April 1." Siri may respond by saying "I don't find a meeting with John on April 1; shall I create it?" Say "Yes" to have Siri create it.

Create Repeating Events

If you want an event to repeat, such as a weekly or monthly appointment, you can set a repeating event.

1. With any view displayed, tap + to add an event.

    The New Event dialog (refer to Figure 22-4) appears.

2. Enter a title and location for the event and set the start and end dates and times, as shown in the previous task, "Add Calendar Events."

3. Scroll down the page if necessary and then tap the Repeat field.

    The Repeat menu shown in **Figure 22-5** is displayed.

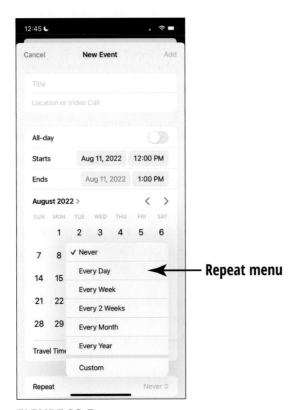

Repeat menu

FIGURE 22-5

4. You have a couple of ways to set an event to repeat:

   - Tap a preset time interval: Every Day, Every Week, Every 2 Weeks, Every Month, or Every Year. You return to the New Event dialog automatically.

   - Tap Custom and make the appropriate settings if you want to set any other interval, such as every two months on the 6th of the month. When finished, tap New Event in the upper left to return to the New Event dialog.

5. To set an expiration date for the repeated event, tap End Repeat and make the necessary settings.

6. Tap Done to return to Calendar.

TIP

Other calendar programs may give you more control over repeating events. If you want a more robust calendar feature, consider setting up your appointments in an application such as the macOS version of Calendar, Outlook, or Google Calendar and syncing it to your iPhone. But if you want to create a simple repeating event in iPhone's Calendar app, simply add the first event on the day of your liking and make it repeat every week. Easy, huh?

View an Event

Tap a Calendar event anywhere — in today, week, month, or list view — to see its details. To make changes to the event you're viewing, tap the Edit button in the upper-right corner.

Add an Alert to an Event

If you want your iPhone to alert you when an event is coming up, you can use the alert feature.

Select a default Calendar Alert

One part of the alert is the sound you'll hear when you're alerted of an upcoming or occurring event. Follow these steps to select a default sound:

1. Tap the Settings icon on the Home screen and choose Sounds & Haptics.

2. Scroll down to Calendar Alerts and tap it; then tap any alert tone which causes iPhone to play the tone for you. Choose the alert tone you want, and then tap Back in the upper-left corner to return to Sounds & Haptics.

TIP

You can set your iPhone to alert you of a Calendar event by using vibration for times when the sound is muted (or the vibration can act as a secondary alert to the alert tones). Tap the Settings icon on the Home screen, tap Sounds & Haptics, tap Calendar Alerts, and then tap Vibration at the top of the screen. Select a vibration pattern from one in the list or create your own by tapping Create New Vibration. You can also disable vibrations for Calendar alerts by tapping None at the very bottom of the screen.

Set up an alert for an event

The other part of an event alert is the text notification you'll see on your iPhone (or other Apple device in which you're signed in to the same iCloud account) when you're alerted of an upcoming or occurring event. Follow these steps to set up this type of alert:

1. Tap the Calendar icon to open the app.

2. Create an event in your calendar or open an existing one for editing, as covered in earlier tasks in this chapter.

3. In the New Event dialog or Edit Event dialog, tap the Alert field.

   The Alert menu appears, as shown in **Figure 22-6**.

4. Tap any preset interval and you'll return to the New Event or Edit Event dialog.

   The Alert setting appears in the dialog, as shown in **Figure 22-7**, and a Second Alert option appears if you'd like to set one.

Alert menu

Alert for an event

FIGURE 22-6

FIGURE 22-7

5. To save all settings, tap Done in the Edit Event dialog or Add in the New Event dialog.

TIP

If you work for an organization that uses a Microsoft Exchange account, you can set up your iPhone to receive and respond to invitations from colleagues. When someone sends an invitation that you accept, it appears on your calendar. Check with your company network administrator (who will jump at the chance to get their hands on your iPhone) or the *iPhone User Guide* (found at https://support.apple.com/guide/iphone/welcome/ios) to set up this feature if it sounds useful to you.

TIP

iCloud offers individuals functionality similar to Microsoft Exchange.

Search for an Event

Can't remember what day next week you scheduled lunch with your brother Mike? You can do a search:

1. With Calendar open in any view, tap the search icon (magnifying glass) in the top-right corner.

2. If the onscreen keyboard doesn't automatically open, tap the search field.

3. Type a word or words to search by and then tap the Search key.

   For example, you might search for *Mike* or *lunch*. While you type, the Results dialog appears.

4. Tap any result to display the event details.

Delete an Event

When an upcoming luncheon or meeting is canceled, you should delete the appointment.

1. With Calendar open, tap an event (see **Figure 22-8**).

2. Tap Delete Event at the bottom of the screen.

3. Tap the Delete Event button again to confirm deletion (see **Figure 22-9**). If this is a repeating event, choose to delete this instance of the event or this and all future instances of the event.

 TIP If an event is moved but not canceled, you don't have to delete the old one and create a new one. Simply change the day and time of the existing event in the Event dialog.

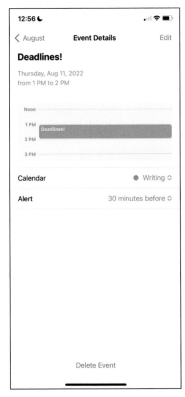

FIGURE 22-8

FIGURE 22-9

Display the Clock App

Clock is a preinstalled app that resides on the Home screen along with other preinstalled apps, such as Books, News, and Camera.

1. Tap the Clock app to open it. If necessary, tap World Clock at the bottom of the screen.

If this is the first time you've opened Clock, the World Clock tab is selected automatically, as shown in **Figure 22-10**. You can add a clock for many locations around the world.

2. Tap + (add) in the upper-right corner.

3. Tap a city in the list. Or tap a letter on the right to display locations that begin with that letter (see **Figure 22-11**), and then tap a city. You can also tap in the search field and begin to type a city name; when the name appears in the results list, tap it.

The new clock appears at the bottom of the World Clock list.

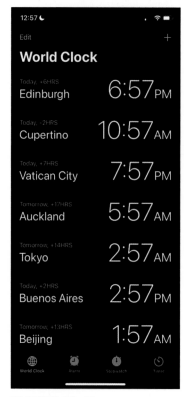

FIGURE 22-10

FIGURE 22-11

Set an Alarm

It seems like nobody has a bedside alarm clock anymore; everyone uses their phone instead. Here's how you set an alarm:

1. With the Clock app displayed, tap the Alarm tab at the bottom of the screen.

2. Tap + (add) in the upper-right corner. In the Add Alarm dialog that appears (see **Figure 22-12**), take any of the following actions, tapping Back after you make each setting to return to the Add Alarm dialog:

- Tap Repeat and repeat the alarm at a regular interval, such as every Monday or every Sunday.

- Tap Label and name the alarm, such as "Take Pill" or "Call Glenn."

- Tap Sound and choose the tune (a ringtone or even a song from your Music app) the alarm will play.

- Tap the Snooze switch on if you want to use the Snooze feature to catch a few extra minutes of sleep.

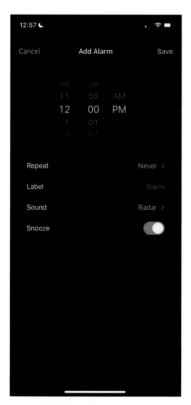

FIGURE 22-12

3. Swipe up or down on the digital clock dials to choose the hour and minute for your alarm. Select the AM or PM button and then tap Save in the upper-right corner.

The alarm appears in the Alarm tab.

TIP

To delete an alarm, tap the Alarm tab at the bottom of the screen and then tap Edit in the upper-left corner. Tap – next to the alarm you want to delete, and then tap the Delete button that appears to the right. The alarm is deleted. Be careful: When you tap the Delete button, the alarm is irretrievable and will need to be re-created from scratch if you mistakenly removed it.

Use Stopwatch and Timer

Sometimes life seems like a countdown or a ticking clock counting the minutes you've spent on a certain activity. You can use the Stopwatch tab in the Clock app to do a countdown to a specific time, such as the moment when your chocolate chip cookies are done cooking. Or use the Timer tab to time an activity, such as a walk.

These two apps work similarly: Tap the Stopwatch or Timer tab from the Clock screen, and then tap the Start button (see **Figure 22-13**).

When you set a timer, iPhone uses a sound to notify you when the time's up. When you start the stopwatch, you have to tap the Stop button when the activity is done.

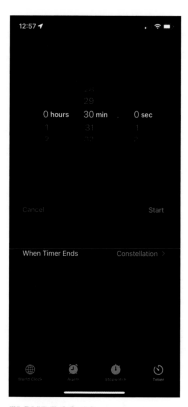

FIGURE 22-13

TIP

Stopwatch allows you to log intermediate timings, such as a lap around a track or the periods of a timed game. With Stopwatch running, just tap the Lap button and the first interval of time is recorded. Tap Lap again to record a second interval, and so forth.

IN THIS CHAPTER

» **Make and edit reminders and lists**

» **Schedule a reminder**

» **Sync reminders and lists**

» **Complete or delete reminders**

» **Set notification types**

» **View notifications and use Notification Center**

» **Set up and turn on a focus**

Chapter **23**

Working with Reminders and Notifications

The Reminders app and Notification Center warm the hearts of those who need help remembering all the details of their lives.

Reminders is a kind of to-do list that lets you create tasks and set reminders so that you don't forget important commitments. You can even be reminded to do things when you arrive at a location, leave it, or receive a message from someone. For example, you can set a reminder so that, when your iPhone detects that you've left the location of your golf game, an alert reminds you to pick up your grandchildren, or when you arrive at your cabin, iPhone reminds you to turn on the water . . . you get the idea. Tags (which were a new feature in iOS 15), such as #groceries or #kids, help you organize and find reminders quickly.

Notification Center allows you to review all the things you should be aware of in one place, such as mail messages, text messages, calendar appointments, and alerts.

If you occasionally need to escape all your obligations, or focus on only certain tasks, try the Focus and Notification Summary features. Turn on these features, and you won't be bothered with alerts and notifications until you're ready to be.

In this chapter, you discover how to set up and view tasks in Reminders and how Notification Center can centralize all your alerts in one easy-to-find place.

Create a Reminder

Creating an event in Reminders is pretty darn simple:

1. Tap Reminders on the Home screen.

2. Tap the New Reminder button in the lower-left corner to add a reminder, as shown in **Figure 23-1**.

   The New Reminder screen appears, together with the onscreen keyboard.

   You can create reminders also from Reminders under My Lists (refer to Figure 23-1). Tap Reminders and then tap the New Reminder button in the lower-left corner of the screen.

3. Enter a task name or description using the onscreen keyboard, and then tap Details to add other items, such as dates, times, and other options (we'll discuss those in the next section).

4. Finally, tap Add in the upper-right corner to create the reminder.

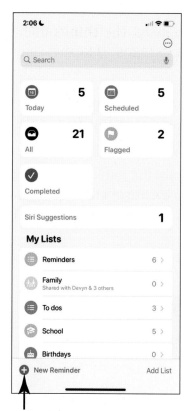

Tap here to add reminders

FIGURE 23-1

Edit Reminder Details

The following task shows you how to add specifics about an event for which you've created a reminder.

1. Tap a reminder and then tap the *i*-in-a-circle (details), which appears to the right of it, to open the Details screen shown in **Figure 23-2**.

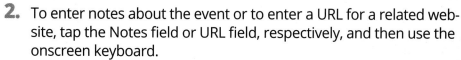 *TIP*

I deal with reminder settings in the following task.

2. To enter notes about the event or to enter a URL for a related website, tap the Notes field or URL field, respectively, and then use the onscreen keyboard.

3. Toggle the Flag switch to enable or disable a flag for the reminder.

   Flags help denote the most important events.

4. Tap Priority and then tap None, Low (!), Medium (!!), or High (!!!) from the choices that appear.

   Priority settings display the number of exclamation points associated with an event in a list to remind you of its importance.

5. Tap List and then tap which list you want the reminder saved to, such as your calendar, iCloud, Exchange, or a category of reminders you've created (see **Figure 23-3**). Tap Details in the upper-left to return to the Details screen.

6. To save changes to the event, tap Done in the upper-right corner.

FIGURE 23-2

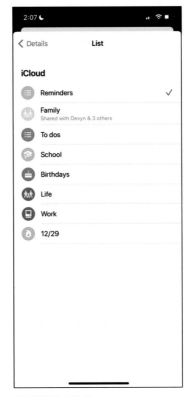

FIGURE 23-3

TIP

For some text fields, Reminders now includes a quick toolbar just above the keyboard to allow you to quickly add a time, a location, tags, a flag, or images to the reminder you've tapped in a list. Just tap the icon for whichever item you want to activate and make the appropriate settings as prompted.

Schedule a Reminder by Time, Location, or When Messaging

One of the major purposes of Reminders is to remind you of upcoming tasks. To set options for a reminder, follow these steps:

1. Tap a task and then tap the *i*-in-a-circle (details) to the right of it.

2. In the dialog that appears (refer to Figure 23-2), toggle the Date switch on (green). In the calendar that appears (see **Figure 23-4**) select a date for your task.

3. Toggle the Time switch on (green) to display the hour and minute dials (see **Figure 23-5**). Use the dials to select a time for the reminder. Select AM or PM as appropriate.

4. If this is something you frequently need to be reminded of, tap Repeat and select an appropriate option. Tap Details to return to the previous screen.

5. Tap the Tags button and add as many tags as you like to your reminder. To add a tag, just type the word using the onscreen keyboard and tap Return. Tap Done in the upper right when finished.

   There's no need to add a hashtag (#) in front of the word for your tag; Reminders automatically adds it.

FIGURE 23-4　　　　**FIGURE 23-5**

6. Toggle the Location switch on and then tap one of the buttons to set a location for your task. Or use the Custom field to enter a location manually, and then tap Details in the upper left to return to the Details screen.

TIP

You have to be in range of a GPS signal for the location reminder to work properly.

7. Scroll down if necessary and then toggle the When Messaging switch on and then tap Choose Person. Select a person or group from your Contacts.

This option will remind you of the item when you're engaged in messaging with the person or group selected. This is a super helpful tool if you, like I, have trouble remembering to share information with people.

8. Add subtasks to this task by tapping the Subtasks option near the bottom of the screen.

9. To attach an image from your photo library, scan a document, or take a photo, tap the Add Image option. Follow the necessary steps based on the option you selected.

Create a List

You can create your own lists of tasks to help you keep different parts of your life organized and even edit the tasks in the list in list view.

1. Tap Reminders on the Home screen to open the app. If a particular list is open, tap Lists in the upper-left corner to return to list view.

2. Tap Add List in the lower-right corner to display the New List form shown in **Figure 23-6**.

3. Tap the List Name text field and enter a name for the list.

4. Tap a color; the list name will appear in that color in list view.

5. Scroll down if necessary, and then tap an icon to customize the icon for the list.

   This feature helps you to better organize your lists by using icons for birthdays, medications, groceries, and host of other occasions and subjects (see **Figure 23-7**).

6. To save the list, tap Done.

Tap an icon to select a custom one for your list

FIGURE 23-6

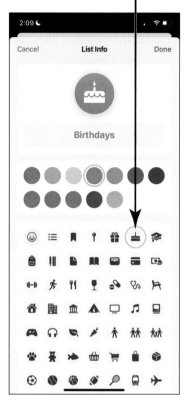

FIGURE 23-7

Sync with Other Devices and Calendars

To make all these settings work, you need to set up your default calendar and enable reminders in your iCloud account.

TIP

Your default Calendar account is also your default Reminders account.

1. To determine which tasks are brought over from other calendars (such as Outlook), tap Settings on the Home screen.

2. Tap your Apple ID, tap iCloud, and then tap the Show All button under the Apps Using iCloud section of the iCloud screen. In the next screen, be sure that Reminders is on (green).

3. Tap iCloud in the upper-left, tap Apple ID in the upper-left on the next screen, and then tap Settings to return to the main Settings screen, swipe up on the screen to scroll down a bit, and then tap Calendar ⇨ Accounts.

4. Tap the account you want to sync Reminders with and then toggle the Reminders switch on, if available (shown in **Figure 23-8**).

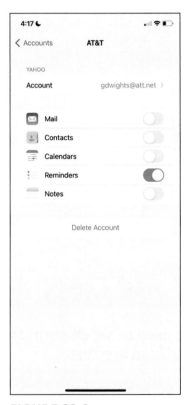

FIGURE 23-8

Mark as Complete or Delete a Reminder

You may want to mark a task as completed or just delete it entirely.

1. With Reminders open and a list of tasks displayed, tap the circle to the left of a task to mark it as complete.

The completed task disappears from the list in a second or two.

2. To view completed tasks, tap the more icon (three dots) in the upper right and then tap Show Completed, as shown in **Figure 23-9**. To hide completed tasks, tap the more icon and then tap Hide Completed.

3. To delete a single task, make sure the list of tasks is displayed and then swipe the task you want to delete to the left. Tap the red Delete button to the right of the task (see **Figure 23-10**) and it will disappear from your list.

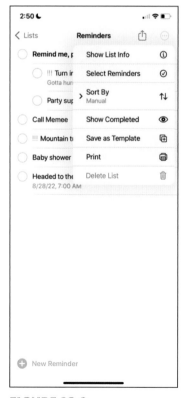

FIGURE 23-9

FIGURE 23-10

WARNING

If you delete a task, it's gone for good. You can't retrieve a deleted task. If you simply want to remove the item from the list without deleting it, mark it as completed, as instructed in Step 1.

4. To delete more than one task, display the list of tasks, tap the more icon (three dots), and then tap Select Reminders. In the screen shown in **Figure 23-11**, tap the circle to the left of each task you want to select, and then tap the delete icon (trash can) at the bottom of the screen.

Tap to select   **Tap to delete**

FIGURE 23-11

Get Notified!

Notification Center is a list of various alerts and scheduled events; it even provides information (such as stock quotes) that you can display by swiping down from the top of your iPhone screen. Notification Center is on by default, but you don't have to include every type of notification there if you don't want to. For example, you may never want to be notified of incoming messages but always want to have reminders listed here — it's up to you.

Notifications are enabled for every app when they're installed, so once you start using your iPhone, you could spend half your day reading or dismissing notifications that you could've waited to see later. To your rescue comes the much-loved Notification Summary feature, which allows you to set up notifications so that you receive them for only some apps as a summary at scheduled times during your day.

Let's jump right in.

Notification summaries

Since notification summaries are such a cool thing, let's take a look at enabling them, and determining which apps are included in the summary.

To enable the Notification Summary feature (if it's not enabled already):

1. Open Settings, tap Notifications, and then tap Scheduled Summary.

2. Toggle the Scheduled Summary switch on (green), as shown in **Figure 23-12**. If this is the first time you're setting this, you'll be walked through several prompts to set up this feature.

3. Adjust your schedules, if you like.

   By default, you get two summaries a day: one at 8 AM and another at 6 PM. From the Schedule section of the Scheduled Summary window, you can

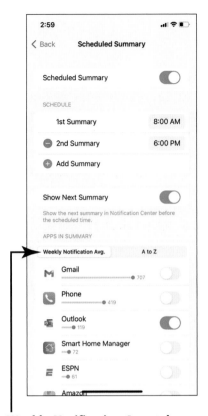

Weekly Notification Avg. tab

FIGURE 23-12

- Tap the + in a green circle (add summary) to add another schedule.
- Tap the − in a red circle (delete), and then tap the red Delete button to delete a schedule.
- Tap a time to the right of a schedule to adjust when the schedule occurs.

4. To add apps to Notification Summary:

(a) Scroll down to the Apps in Summary section of the Scheduled Summary screen.

(b) Tap the Weekly Notifications Avg. tab (labeled in Figure 23-12) or the A to Z tab to see a list of apps by an average of how many notifications you receive from them or by alphabetical order, respectively.

(c) Toggle the switch on (green) for each app you want to appear in the notification summary.

Note the line with the red dot below each app in the Weekly Notifications Avg. tab. You can't do anything with that line or dot; it's only an indicator of the daily average of notifications that the app generates.

TIP

You can feel comfortable adding all your apps to Notifications Summary, if you like. Time-sensitive messages, such as phone calls and texts, will break through anyway.

Set notification types

Some Notification Center settings let you control what types of notifications are included:

1. Tap Settings and then tap Notifications.

The Notification Style section lists the apps included in Notification Center. The app's state is listed directly under its name. For example, *Immediate* appears below AccuWeather in **Figure 23-13**, indicating the method of notifications enabled for that app.

2. Tap any app to open its settings.

3. Set an app's Allow Notifications switch (see **Figure 23-14**) on (green) or off, to include or exclude it, respectively, from Notification Center.

4. In the Notification Delivery section, tap Immediate Delivery (to receive notifications for this app immediately) or Scheduled Summary (to add the app to the notification summary; see the preceding task in this chapter for more info).

TIP

If you select Scheduled Summary, the Alerts section is no longer available.

5. In the Alerts section, you can choose to display alerts on the lock screen, in Notification Center, as banners, or as a combination.

If you don't want any alerts, simply don't make a selection.

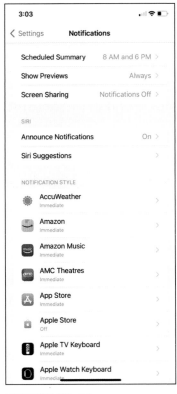

FIGURE 23-13

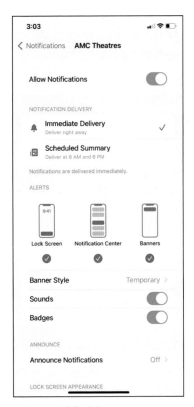

FIGURE 23-14

TIP

6. If you enabled Banners, choose a style by tapping the Banner Style option. Tap the name of the app in the upper left to return to the previous screen.

Banners will appear and then disappear automatically if you tap the Temporary style. If you choose Persistent, you have to take an action to dismiss the alert when it appears (such as swiping it up to dismiss it or tapping to view it).

7. Toggle the Sounds and Badges switches to suit your needs.

8. Scroll down and tap Show Previews to determine when or if previews of notifications should be displayed on your iPhone's screen.

Options are Always (the default), When Unlocked (previews appear only when your iPhone is unlocked), or Never. Tap the name of the app to go to the previous screen.

9. Select a Notification Grouping option. Then tap the Back button or the name of the app to return to the previous screen.

This feature enables you to group notifications if you like, as opposed to seeing every single notification listed. Options are

- *Automatic:* Notifications are grouped according to their originating app but may also be sorted based on various criteria. For example, you may see more than one group for Mail if you receive multiple emails from an individual; those email notifications may merit their own grouping.

- *By App:* Notifications are grouped according to their originating app — period. You'll see only one grouping for the app, not multiple groups based on the varying criteria, as described for the Automatic setting.

- *Off:* All notifications for this app will be listed individually.

10. Tap Notifications in the upper-left corner to return to the main Notifications settings screen. When you've finished making settings, press the Home button or swipe up from the bottom of the screen (iPhone models without a Home button).

View Notification Center

After you've made settings for what should appear in Notification Center, you'll want to look at those alerts and reminders regularly.

1. Swipe down from the top of any screen to display Notification Center (see **Figure 23-15**).

If your iPhone doesn't have a Home button, be sure to swipe from the top center so you don't accidentally open Control Center.

2. To close Notification Center, swipe upward from the bottom of the screen.

TIP

To determine what is displayed in Notification Center, see the preceding task.

Notification Center has two sections for you to play with: Notification Center and Today.

1. Swipe down from the top of the screen to open Notification Center.

Notifications are displayed by default.

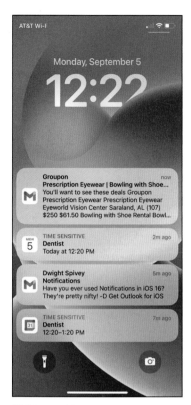

FIGURE 23-15

2. Swipe from left to right on the date/time at the top of Notification Center to access the Today section.

Here you can view information in widgets that pertain to today, such as reminders, weather, stock prices, calendar items, and other items you've selected to display in Notification Center (see the preceding task).

TIP

You can select which widgets appear on the Today screen. From the first Home screen, swipe from left to right to access the Today screen, tap the Edit button, and then select the items you want to see. Tap Done to return to finish customizing your Today screen.

3. Swipe from right to left anywhere in the Today screen to go back to the Notifications section to see all notifications that you set up in the Settings app.

You'll see only notifications that you haven't responded to, haven't deleted in the Notifications section, or haven't viewed in their originating app.

Stay Focused and Undisturbed

The Focus feature is really an extension of (and incorporates) the ever-popular Do Not Disturb feature.

Focus keeps you from being disturbed by incoming calls and notifications during various times and tasks. You can customize a list of people or apps that can still contact or notify you, even when a focus is enabled. When you turn on a focus for your iPhone, it's automatically turned on for every other Apple device that you're signed into using the same Apple ID.

Do Not Disturb is a simple but useful setting you can use to stop alerts, phone calls, text messages, and FaceTime calls from appearing or making a sound. You can make settings to allow calls from certain people or several repeat calls from the same person in a short time to come through. (The assumption here is that such repeat calls may signal an emergency or an urgent need to get through to you.)

REMEMBER

I discussed these features as they pertain to driving in Chapter 6. I focus on (pun intended) other aspects of them here.

Set up a focus

To set up a focus:

1. Go to Settings and tap Focus.

2. Tap either Do Not Disturb or the particular focus you'd like to edit.

3. In the Allow Notifications section (see **Figure 23-16**), tap People or Apps to customize who or what app can contact or notify you, even when Do Not Disturb or the particular focus is on.

4. Tap the Add People or Add Apps button.

5. Tap the names of people or apps for which you want to allow exceptions, and then tap Done in the upper-right corner.

   The Allowed People or Allowed Apps area displays the people or apps for which you've allowed exceptions, as shown in **Figure 23-17**.

FIGURE 23-16

FIGURE 23-17

6. Tap – in the upper-left corner of an icon to remove an individual person or app from the list.

7. To exit, tap the button in the upper-left corner.

8. If you prefer it to turn on the focus at a specific time, use the options in the Turn On Automatically section to set a schedule.

Turn on a focus

To turn on a focus:

1. Open Control Center.

   For iPhones with Home buttons, swipe up from the very bottom of your screen. For iPhones without a Home button, swipe down from the upper-right corner of your screen.

2. Tap the text on the focus button.

3. Tap a focus in the list to turn it on.

   When a focus is on, its button is white, as shown in **Figure 23-18**.

4. Tap the more icon (three dots) to the right of a focus name to access more options, as shown in **Figure 23-19**.

   These options may vary, depending on the focus.

5. To turn off a focus, open Control Center, tap the text on the focus button, and then tap the name of the enabled focus. Once disabled, the button for the focus turns dark gray.

FIGURE 23-18

FIGURE 23-19

Chapter **24**

Keeping Tabs on Your Health

A pple has begun a major push into health and wellness, and the Health app is an important tool in its quest to help customers achieve a healthier lifestyle. Essentially, Health is an aggregator for health and fitness data, and it can be accessed from the first Home screen on your iPhone. You can input information about your height, weight, medications, nutrition, sleep quality, and more, either manually or automatically through data collected by your iPhone or Apple Watch. You can then view health-related statistics based on that data in the dashboard view.

In this chapter, I provide an overview of the Health app and show you how to get information into and out of it. I also give you a glimpse of some of the apps and health equipment that interact with Health to make it even more useful in months and years to come.

Understand the Health App

The Health app is meant to be a one-stop repository for your health information. You can manually input information, and the app can also collect health data from other health-related apps and equipment that support working with the Health app, such as an Apple Watch.

REMEMBER

Check with the app's developer or the equipment's manufacturer for details on whether their product supports and works with the Health app in iOS.

The first time you open Health, you'll see screens informing you of what's new in iOS 16. As you progress through those screens, the last one will allow you to set up your Health profile. You can also set up notifications for Cardio Fitness, Walking Steadiness, and Trends.

If you didn't set up your profile at that time, you can do so now:

1. Open the Health app by tapping its icon on the Home screen.

2. Tap the Apple ID button (your picture or initials in a circle) in the upper-right corner.

3. Tap Health Details to see the profile page shown in **Figure 24-1**.

   This page contains the most basic information about you, such as your date of birth, sex, height, and weight.

4. Tap the blue Edit button.

5. Tap a field and enter the appropriate information. Repeat with other fields as necessary.

6. When you're finished, tap the blue Done button, tap Profile in the upper left, and then tap the Done button again.

TIP

Information entered into your Health Profile is crucial to the accuracy of health-related apps and functions, such as the number of calories burned when using the Workout app on your Apple Watch. Try to be as accurate (and honest) as possible when filling out the info.

The next stop is the Summary screen, shown in **Figure 24-2**. Summary is a snapshot of health-related highlights and metrics you've accumulated, and may include alerts and messages.

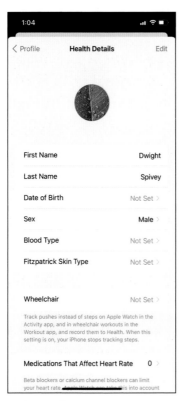

Apple ID button

FIGURE 24-1

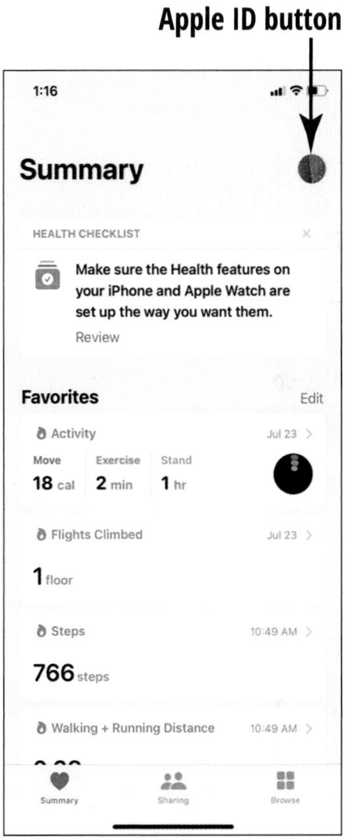

FIGURE 24-2

The Favorites section is a list of metrics that have been collected about your health through apps and devices. You can customize the metrics that are displayed in this section:

1. Tap the blue Edit button in the upper-right corner of the Favorites section.

2. Tap the star next to metrics you want to add (the star turns solid blue) or remove (the star turns white), as shown in **Figure 24-3**.

If you don't see a metric you want to add, tap the All tab at the top to reveal tons of other hidden options that you can select.

3. When you're finished, tap the Done button in the upper right.

The Favorites section displays your customized content.

Tap on any of the metrics in the Favorites section to see a much more detailed breakdown of the item or activity.

Health is divided into categories such as Activity, Mindfulness, Mobility, Nutrition, Sleep, Symptoms, and Other Data (such as blood glucose and the number of times fallen), as shown in **Figure 24-4**. You see these categories when you tap the Browse button at the bottom of the Health screen.

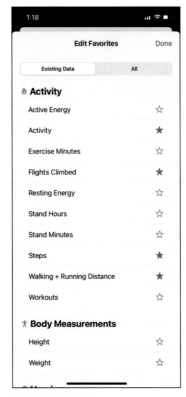

FIGURE 24-3

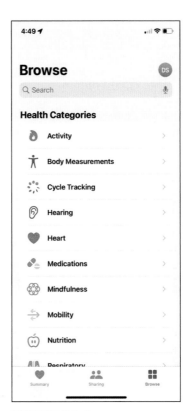

FIGURE 24-4

Each category can have subtopics, which you can reach by tapping the category, such as Hearing (shown in **Figure 24-5**).

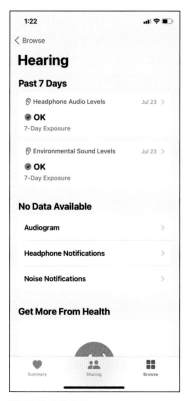

FIGURE 24-5

The Health app also has a medical record feature called Medical ID, which is covered in the next task, as well as the capability to help you sign up to be an organ donor.

TECHNICAL STUFF

At some point during your use of the Health app, you may be asked to share your health and activity data with Apple. Whether you do so is up to you, but be assured that if you elect to share that data with Apple, it will be done so anonymously and confidentially. None of your personal information will be shared as part of the health and activity data.

Apps that Health can collaborate with

Health is essentially an information aggregator, and it is continuing to grow in terms of available apps designed to interact with it to supply imported data, such as calories consumed and steps walked. It's essential to have this aggregator so that you don't have to open 20 different apps to keep track of how you're doing health-wise. Instead you can find it all in one convenient location.

Here are just a handful of the apps that work with the Health app: Aaptiv: #1 Audio Fitness App, Calm, MyPlate Calorie Counter, Medisafe, Dexcom G6 Mobile, Lose it!, Sleep Cycle, Zova, Mayo Clinic, MyChart, Nike apps, Fitbit, CARROT Hunger, Human — Activity Tracker, and MyFitnessPal.

Equipment that connects with Health

Health is designed to connect with a variety of equipment to wirelessly import data (via Bluetooth and Wi-Fi) about your health and fitness. This equipment includes treadmills, cycling computers, indoor bike trainers, pulse oximeters, electronic toothbrushes, thermometers, glucose monitors, sleep monitors, posture trainers, smart jump ropes (no kidding!), scales, EKGs or ECGs (electrocardiograms), Apple Watch (www.apple.com/watch), blood-pressure monitors, and a wide variety of more devices, with more on the way seemingly every day.

You can learn more about the Health app and find the latest related news by visiting Apple's Health website at www.apple.com/ios/health.

Create Your Medical ID

One of the simplest Health features is Medical ID, which allows you to store your vital statistics. This could be useful if you're in an accident and emergency medical personnel need to access your blood type or allergies to medications.

1. In the Health app, tap the Apple ID button in the upper-right corner (refer to Figure 24-2), and then tap the Medical ID option.

2. Tap Get Started on the Set Up Your Medical ID screen.

3. Scroll down and tap the + next to items on the screen (such as blood type) to add that particular information to your ID.

TIP

   Be as specific (and as honest) as possible when entering your medical information! The more info healthcare providers have, the better.

4. Tap a field, such as Medical Conditions or Allergies & Reactions, and enter information using the onscreen keyboard that appears.

5. Toggle the Show when Locked switch on (green) to allow your Medical ID to be accessed from the lock screen.

   This feature is important to turn on so that emergency responders can see your medical information without needing you to allow them access.

6. Toggle the Share During Emergency Call switch on to allow your Medical ID information to be sent to emergency services when you call or text them if you're in a part of the world that supports this feature.

   This feature gives emergency responders as much information as possible about you before they arrive so that they can be better prepared to assist.

7. To save your entries, tap Done in the upper-right corner.

Become an Organ Donor

Apple is the first company to make it easy to sign up as an organ donor through its operating system (iOS, the software that controls your iPhone). To sign up to be an organ donor, follow these steps:

1. In the Health app, tap the Apple ID button in the upper-right corner (refer to Figure 24-2).

2. Under Features, tap the Organ Donation option.

3. Tap Sign Up with Donate Life to become a donor. If you instead want to find out more about being an organ donor, tap Learn More.

4. On the next screen, enter your information as needed and then tap Continue at the bottom of the page.

5. When asked to confirm your registration with Donate Life, tap the Complete Registration with Donate Life button at the bottom of the screen to confirm.

View Health App Data

Knowing where and how to view your health data in the Health app is important if you want to get the most from its features.

1. Tap the Browse button at the bottom of the Health app screen.

2. Tap a category, such as Activity.

3. Tap one of the subtopics, such as Steps, and you'll see a screen similar to **Figure 24-6**.

TIP

Your iPhone provides a lot of information to Health, but you can really step it up a notch by wearing an Apple Watch! The Apple Watch pairs (connects wirelessly) with your iPhone and shares with the Health app the information it collects through its multitude of sensors, such as heart rate information (resting rate and variability). That information is collected and saved for every day you wear the watch.

This screen graphically displays data about your activity daily, weekly, monthly, every six months, or yearly (tap D, W, M, 6M, or Y in the top bar, respectively). As you continue to scroll down, you'll see the activity data broken down into even more helpful categories. You'll find a description of how this subtopic is monitored and why, and a list of apps related to the subtopic, like the one shown in **Figure 24-7**.

4. Scroll all the way down to the bottom of the screen and tap Data Sources & Access.

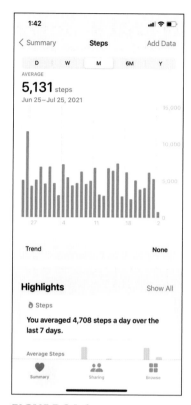

FIGURE 24-6

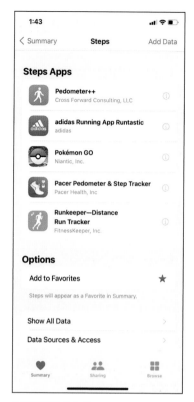

FIGURE 24-7

If you have other health-related apps on your iPhone that can receive data from Health, they'll appear in the Apps Allowed to Read Data section.

5. To enable or disable apps that can receive data from Health, tap Edit in the upper-right corner, tap the app's name to enable it (check mark) or disable it, and then tap Done in the upper right.

6. In the Data Sources section, tap a device name to see the data it is reporting to Health.

7. Tap Back in the upper-left corner of the screen to return to the previous screen, and then, in the upper left, tap the name of the subtopic you selected in Step 3.

8. Tap Show All Data near the bottom of the screen to see a table of entries, including your most recent entry.

Import and Export Health Data

Some apps from which you can import data are MyFitnessPal; Strava: Run, Ride, Swim; and LifeSum. Different apps may send or receive data from Health using different interfaces and commands. But, in essence, here's how that process works: When you use an app that supports Health, the app requests permission to update data, which saves you the drudgery of entering the data manually.

Here's how you view supporting apps that you've downloaded and installed on your iPhone:

1. In the Health app, tap the Apple ID button in the upper-right corner.

2. In the Privacy section, tap Apps.

3. Tap each app to find out and modify what data it supplies to the Health app.

In addition, you can export data using such apps as the Mayo Clinic app so that you can keep your physician informed about your progress or challenges.

TIP

You can also export all of your Health data as a file that you can provide to caregivers. Tap the Apple ID button in the upper right, tap the Export All Health Data button at the bottom of the next screen, tap Export to confirm, and then select a method for providing the information (text message, email, AirDrop, and more). This file may be large if you've saved a lot of information in Health, so saving it to and sharing it from the cloud using iCloud or another service such as Google Drive or Dropbox may be preferred.

Sharing Is Caring

iOS 16 is great at sharing, and the Health app is one of those apps that utilizes this virtue. Sharing Health data is different than exporting a file containing your Health info and giving it to someone, such as a doctor. Sharing is allowing those you trust to have a real-time view of your Health app, so they can see how things are going at any moment. Those you choose to share with will receive notifications regarding your Health info; it's great to have someone else in your corner. New to iOS 16 is the ability to invite others to share their health information with you. When they receive your invitation, they'll be able to decide what information to share with you.

Please know that you're in total control of who you share with and what information you share. You can also share certain information with a specific person that you don't share with anyone else. And you can stop sharing any time you please.

TECHNICAL STUFF

The people you're sharing with must be running iOS 15 or newer on their iPhone. This feature may not be available in some countries.

To share Health data with others:

1. In the Health app, tap the Sharing tab at the bottom of the screen.
2. Tap the blue Share with Someone button.
3. Enter the name of someone in your contacts that you want to share with, and then tap the name.
4. Follow the prompts on the next several screens to determine what information you'd like to share.
5. When you reach the Here's What You've Chosen to Share screen, tap the blue Share button to send an invitation to the person you're sharing with.
6. When you're finished, tap Done.

To stop sharing Health data:

1. Tap the Sharing tab on the Health app screen.
2. Tap the name of the person you want to stop sharing with.
3. Scroll down to the bottom of the page, and tap the red Stop Sharing button.

View Health Records

Apple's foray into healthcare is gaining momentum rapidly as patients, doctors, and other providers rely more heavily on the Apple ecosystem of devices and software. Today, hundreds of healthcare institutions are using the Health app to allow patients to view their health records on their iPhone. (Go to `https://support.apple.com/en-us/HT208647` to see the ever-growing list of participants.)

Contact your health provider and make sure they provide information through the Health app. Once you confirm that, ask them what account information you need to connect to their systems so that you can view your records. You'll use this information in Step 5.

1. Tap the Browse button at the bottom of the Health app screen.

2. Scroll down the screen until you reach the Health Records section and tap the Add Account button.

3. Tap Allow While Using App or Allow Once to allow the Health app to access your location.

   This feature helps the app locate providers in your area that currently provide health records through the Health app.

4. In the list that appears, locate and tap your provider, and then tap the Connect to Account button.

5. From this point, enter the account information you gathered from your provider.

   Now you can use the Health app to view your health records from that provider.

Stay on Top of Your Medications

I think one of the best new additions to iOS 16 is the ability to track your medications and set up schedules and reminders for them, right from the Health app. To get started:

1. Tap the Browse button at the bottom of the Health app screen, and then tap the Medications button.

2. On the Medications screen, tap the Add a Medication button.

3. On the Add Medication screen, tap the search field to the find the medication. Or use your iPhone's camera to scan your medication bottle for information.

4. Follow the onscreen instructions to find the correct information for your medication and dosages, tapping the Next button after every selection.

5. Create your schedule for taking your medications.

6. If you are prompted to choose the shape of your medication, select a shape or click the Skip button.

7. Select the color of your medication and a background color.

   If you don't see an exact color match for the medication, select the closest thing you can find. For the background color, a contrasting color is a good idea.

8. If you want the Health app to alert you to any potential interactions with substances such alcohol, cannabis, or tobacco, toggle the switch for each one. Then tap Done.

   Your medication will appear on the Medications screen, as shown in **Figure 24-8**.

9. Repeat Steps 2 through 8 for each medication you want to add to the Health app.

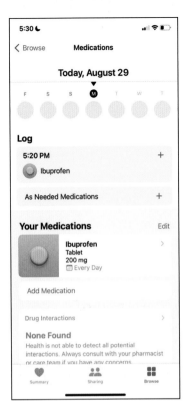

FIGURE 24-8

IN THIS CHAPTER

» Clean and protect your iPhone

» Extend your iPhone's battery life

» Fix a problematic iPhone

» Update iOS

» Find a missing iPhone

» Back up your iPhone

Chapter **25**

Troubleshooting and Maintaining Your iPhone

Phones don't grow on trees — in fact, they cost a pretty penny. That's why you should learn how to take care of your iPhone and troubleshoot many of the problems it might have so that you get the most out of it.

In this chapter, I provide some advice about the care and maintenance of your iPhone, as well as tips for solving common problems, updating iPhone system software (iOS), and even resetting the iPhone if something goes seriously wrong. In case you lose your iPhone, I even tell you about a feature that helps you find it, activate it remotely, or even disable it if it has fallen into the wrong hands (which is highly important if you're an international spy or the like, or you simply don't want someone messing with your stuff). Finally, you get information about backing up your iPhone settings and content using iCloud.

Keep the iPhone Screen Clean

If you've been playing with your iPhone, you know that it's a fingerprint magnet (even though Apple claims that the iPhone has a fingerprint-resistant screen). Here are some tips for cleaning your iPhone screen:

» **Use a dry, soft cloth.** You can get most fingerprints off with a dry, soft cloth, such as the one you use to clean your eyeglasses or a cleaning tissue that's lint and chemical free. Or try products used to clean lenses in labs, such as Kimwipes (which you can get from several major retailers, such as Amazon, Walmart, and office supply stores).

» **Use a slightly dampened soft cloth.** This may sound counter-intuitive to the preceding tip, but to get rid of more stubborn grime, very (and I stress, very) slightly dampen the soft cloth. Again, make sure that whatever cloth material you use is free of lint.

» **Remove the cables.** Turn off your iPhone and unplug any cables from it before cleaning the screen with a moistened cloth, even a very slightly moistened one.

» **Avoid too much moisture.** It's best to keep your phone dry. iPhone 7 and later models have some degree of water resistance, but you don't want to press your luck. So you don't have to freak out if you splash some water on your iPhone, but do be sure to wipe it dry right away, and it should be fine. However, you shouldn't expect it to be hunky-dory after it sinks to the bottom of Lake Erie (or just any other lake, for that matter).

» **Don't use your fingers!** That's right. By using a stylus rather than your finger, you avoid smearing oil from your skin or cheese from your pizza on the screen. A number of top-notch styluses are out there; just search Amazon for *iPhone stylus* and you'll be greeted with a multitude of them (most are reasonably priced).

Apple's Pencil unfortunately does not work with iPhones and is only compatible with iPads. Don't mistake the Apple Pencil for an iPhone stylus.

» **Never use household cleaners.** They can degrade the coating that keeps the iPhone screen from absorbing oil from your fingers. Plus, there's just simply no need to go that far since the screen cleans quite easily with little or no moisture.

Don't use premoistened lens-cleaning tissues to clean your iPhone screen! Most brands of wipes contain alcohol, which can damage the screen's coating.

Protect Your Gadget with a Case

Your screen isn't the only element on the iPhone that can be damaged, so consider getting a case for your iPhone so that you can carry it around the house or travel with it safely. Besides providing a bit of padding if you drop the device, a case makes the iPhone less slippery in your hands, offering a better grip when working with it.

Several types of covers and cases are available, but be sure to get one that will fit your model of iPhone because their dimensions, thickness, and button placements may differ. There are differences between covers and cases:

» **Covers tend to be more for decoration than overall protection.** While they do provide some minimal protection, they're generally thin and not well-padded.

» **Cases are more solid and protect most, if not all, of your iPhone.** They're usually a bit bulky and provide more padding than covers. If you're prone to dropping things, a good case is by far your best bet.

Extend Your iPhone's Battery Life

The much-touted battery life of the iPhone is a wonderful feature, but you can do some things to extend it even further. Here are a few tips to consider:

>> **Keep tabs on remaining battery life.** You can estimate the amount of remaining battery life by looking at the battery icon on the far-right end of the status bar, at the top of your screen. Swipe down from the upper right (or swipe up from the bottom if your iPhone has a Home button) to open Control Center, which shows the battery percentage remaining.

>> **Wired charging is generally faster than wireless charging.** Charging your iPhone using the Lightning-to-USB (or USB-C) cable and a power adapter is faster than using a wireless charging pad. For example, an iPhone 8, when plugged in with a USB-C charger, can draw up to 18 watts and charge to 50 percent in about 30 minutes. But when you set an iPhone 8 on a wireless charging pad, it can draw only 7.5 watts and will charge more slowly. An iPhone 12 can draw up to 20 watts from a plugged-in USB-C charger, but with a fancy new MagSafe wireless charger, it's limited to 15 watts. So if you're in a hurry and need a quick charge, plugging in is the way to go.

TIP

If you're charging your iPhone overnight, you don't need fast charging. Plugging it in with an old 5-watt charger is perfectly fine, as is using a 7.5-watt wireless charging pad. Some believe that slower charging is better for overall battery health because your phone doesn't get as hot during the charging process.

>> **Use a case with an external battery pack.** These cases are handy when you're traveling or unable to reach an electrical outlet easily. However, they're also a bit bulky and can be cumbersome in smaller hands.

>> **If your battery charge is low, consider turning off wireless connectivity options that you may not be using.** For example, if you're not near a Wi-Fi connection, temporarily turn off Wi-Fi.

» **Dim the screen, or use auto-brightness, which automatically adjusts the brightness of your screen according to external lighting conditions.** Auto-brightness can be enabled or disabled by toggling its switch on (green) or off (gray) at Settings ⇨ Accessibility ⇨ Display & Text Size ⇨ Auto-Brightness.

» **The fastest way to charge your iPhone is to turn it off while charging it.** If turning off your iPhone doesn't sound like the best idea for you, you can disable Wi-Fi or Bluetooth (or preferably both) to facilitate a faster recharge. Also, charging will take longer if you use your iPhone while charging it.

If you activate airplane mode to turn off both Wi-Fi and Bluetooth at the same time, remember that this also disables your cellular connection. You won't be able to place or receive calls until you deactivate airplane mode. A better alternative is to simply turn off Wi-Fi and Bluetooth individually, bypassing airplane mode altogether (unless you're instructed to activate it while on an airplane, of course).

» **The battery icon on the Status bar indicates when the charging is complete.**

Be careful not to use your iPhone in ambient temperatures higher than 95 degrees Fahrenheit (35 degrees Celsius), as doing so may damage your battery. Damage of this kind may also not be covered under warranty. Charging in high temperatures may damage the battery even more.

If you notice that your battery won't charge more than 80 percent, it could be getting too warm. Unplug the iPhone from the charger and try again after the phone has cooled down a bit.

Your iPhone battery is sealed in the unit, so you can't replace it yourself the way you can with many laptops or other cellphones. If your iPhone is out of warranty, you have to fork over upwards of $69 to have Apple install a new one.

For much more detailed information on your iPhone's battery, I suggest visiting Apple's iPhone Battery and Performance website at `https://support.apple.com/en-us/HT208387`.

CHECK YOUR BATTERY'S STATISTICS AND HEALTH

iOS 16 has a great way to keep track of your iPhone's battery health and how it's being used. Go to Settings ⇨ Battery to view stats about your battery over the last 24 hours (see the figure) and last 10 days. You can also enable low power mode, which temporarily disables some non-essential activities that may zap your battery's charge until you're able to give it some more juice. Tap the Battery Health button to view information on how much lifespan your battery may possess.

Deal with a Nonresponsive iPhone

If your iPhone goes dead on you, it's most likely a power issue, so the first thing to do is to plug the Lightning-to-USB or Lighting-to-USB-C cable into the USB or USB-C power adapter, respectively, plug the power adapter into a wall outlet, plug the other end of the cable into your iPhone, and charge the battery.

Another thing to try — if you believe that an app is hanging up the iPhone — is to double-click the Home button until App Switcher appears, and then swipe up on the cantankerous app. For iPhones without a Home button, swipe up from the bottom of the screen, pause and hold your finger on the screen until App Switcher opens, and then swipe up on the app that's causing issues. The problematic app should then close.

You can always use the tried-and-true reboot procedure: On iPhones with a Home button, press the sleep/wake button (on iPhones without a Home button, press both the side and volume up buttons) until the Slide to Power Off slider appears. Drag the slider to the right to turn off your iPhone. After a few moments, press the sleep/wake or side button to boot up the little guy again.

If the situation seems drastic and none of these ideas works, try to force restart your iPhone. To do this, quickly press and release the volume up button, quickly press and release the volume down button, and then press and hold down the side button until the Apple logo appears onscreen.

TIP
If your phone has this problem often, try closing out some active apps that may be running in the background and using up too much memory. To do this, press the Home button twice to open App Switcher (swipe up from the bottom of the screen and pause for iPhones without a Home button) and then from the screen displaying active apps, swipe an app upward. Also check to see that you haven't loaded up your iPhone with too much content, such as videos, which could be slowing down its performance.

Update the iOS Software

Apple occasionally updates the iPhone system software, known as iOS, to fix problems or offer enhanced features. You should occasionally check for an updated version (say, every month). You can check by connecting your iPhone to a recognized computer (that is, a computer you've used to sign into your Apple account before) with iTunes installed, but it's even easier to just update from your iPhone Settings, though it's a tad slower:

1. Tap Settings from the Home screen.

2. Tap General and then tap Software Update (see **Figure 25-1**).

   A message tells you whether your software is up-to-date.

3. If your software is not up-to-date, tap Download and Install and follow the prompts to update to the latest iOS version.

FIGURE 25-1

You can also allow your iPhone to automatically update iOS. Go to Settings⇨General⇨Software Update⇨Automatic Updates, then toggle both the Download iOS Updates switch and the Install iOS Updates switch on (green).

Find a Missing Apple Device

The Find My app can pinpoint the location of your Apple devices and your Apple-using friends. This app is extremely handy if you forget where you left your iPhone or someone absconds with it. Find My not only lets you track down the critter but also lets you wipe out the data contained in it if you have no way to get the iPhone (or other Apple device) back.

You must have an iCloud account to use Find My. If you don't have an iCloud account, see Chapter 4 to find out how to set one up.

If you're using Family Sharing, someone in your family can find your device and play a sound on it. This works even if the volume on the device is turned down.

Follow these steps to set up the Find My feature for your iPhone:

1. Tap Settings on the Home screen.

2. In Settings, tap your Apple ID at the top of the screen and then tap Find My.

3. In the Find My settings, tap Find My iPhone and then tap the on/off switch for Find My iPhone to turn the feature on (see **Figure 25-2**).

You may also want to turn on the Find My Network option. This allows Apple devices to be found using their built-in Bluetooth technology, even when not connected to Wi-Fi or a cellular network. When you mark your device as missing on www.icloud.com and another Apple user is close by the device, the two devices connect anonymously via Bluetooth and you're notified of its location. Pretty cool stuff, and completely private for all involved parties.

FIGURE 25-2

TIP

You may also want to enable the Send Last Location switch to allow your iPhone to send its location to Apple when the phone's battery is running low.

From now on, if your iPhone is lost or stolen, you can go to www.icloud.com from your computer, your iPad, or another iPhone and enter your Apple ID and password. You can also use the Find My app on your iPhone, iPad, or Mac.

4. In your computer's browser, the iCloud Launchpad screen appears. Click the Find iPhone button to display a map of your device's location and some helpful tools.

5. Click the All Devices option at the top of the window and click your iPhone in the list. In the window that appears, choose one of three options:

- *Play Sound:* Your iPhone plays a ping sound that might help you locate it if you're in its vicinity.

- *Lost Mode:* Locks the iPhone so others can't access it. You can send a note to whoever has your iPhone with details on how to return it to you.

- *Erase iPhone:* Wipes information from the iPhone.

WARNING

The Erase iPhone option deletes all data from your iPhone, including contact information and content (such as music). However, even after you've erased your iPhone, it will display your phone number on the lock screen along with a message so that any Good Samaritan who finds it can contact you. If you've created an iTunes or iCloud backup, you can restore your iPhone's contents from those sources.

Back Up to iCloud

You used to be able to back up your iPhone content using only iTunes, but since Apple's introduction of iCloud, you can back up via a Wi-Fi network to your iCloud storage. You get 5GB of storage for free. You can pay for increased storage: a total of 50GB for $0.99 per month, 200GB for $2.99 per month, or 2TB for $9.99 per month.

You must have an iCloud account to back up to iCloud. If you don't have an iCloud account, see Chapter 4 to find out more.

To perform a backup to iCloud:

1. Tap Settings from the Home screen and then tap your Apple ID at the top of the screen.

2. Tap iCloud and then tap iCloud Backup.

3. In the Backup Your iPhone with iCloud pane that appears, tap the Back Up This iPhone switch to enable automatic backups. To perform a manual backup, tap Back Up Now, which will appear after you've enabled automatic backup.

   A progress bar shows how your backup is moving along.

If you get your iPhone back after it wanders and you've erased it, just enter your Apple ID and password and you can reactivate it.

TIP

You can also back up your iPhone using iTunes (for Macs running macOS Mojave or earlier, and for Windows-based PCs) or Finder (for Macs running macOS Catalina or newer). This method saves more types of content than an iCloud backup, and if you have encryption turned on in iTunes or Finder, it can save your passwords as well. However, this method requires you to connect the iPhone to a computer to perform the backup. If you do back up and get a new iPhone down the line, you can restore all your data to the new phone easily.

Index

About the Author

Dwight Spivey has been a technical author and editor for over a decade, but he's been a bona fide technophile for more than three of them. He's the author of *Apple Watch For Seniors For Dummies* (Wiley), *iPad For Seniors For Dummies* (Wiley), *Idiot's Guide to Apple Watch* (Alpha), *Home Automation For Dummies* (Wiley), *How to Do Everything Pages, Keynote & Numbers* (McGraw-Hill), and many more books covering the tech gamut.

Dwight is also the Director of Educational Technology at Spring Hill College. His technology experience is extensive, consisting of macOS, iOS, Android, Linux, and Windows operating systems in general, educational technology, learning management systems, desktop publishing software, laser printers and drivers, color and color management, and networking.

Dwight lives on the Gulf Coast of Alabama with his wife, Cindy, their four children, Victoria, Devyn, Emi, and Reid, and their pets Rocky, Penny, and Mirri.

Dedication

To my precious niece, Kamilla! The sun shines just a little bit brighter when you smile. Love you, sweet girl!

Author's Acknowledgments

Carole Jelen, my long-time agent; you're always first in line for my deepest thanks!

Tremendous gratitude goes to Elizabeth Stilwell, Susan Pink, and Guy Hart-Davis. And of course, the editors, designers, and other wonderful Wiley professionals who are absolutely critical to the completion of these books I'm so blessed to write. I sincerely want every individual involved at every level to know how appreciative I am of their dedication, hard work, and patience in putting together this book.

Publisher's Acknowledgments

Associate Editor: Elizabeth Stilwell

Project Editor: Susan Pink

Production Editor:
  Saikarthick Kumarasamy

Technical Reviewer: Guy Hart-Davis

Proofreader: Debbye Butler

Cover Image:
  © Olga Breslavtsev/Getty Images